The Young Hyperactive Child: Answers to Questions About Diagnosis, Prognosis, and Treatment

The *Journal of Children in Contemporary Society* series:

- *Young Children in a Computerized Environment*
- *Primary Prevention for Children & Families*
- *The Puzzling Child: From Recognition to Treatment*
- *Children of Exceptional Parents*
- *Childhood Depression*
- *Newcomers to the United States: Children and Families*
- *Child Care: Emerging Legal Issues*
- *A Child's Brain: The Impact of Advanced Research on Cognitive and Social Behaviors*
- *Teachers: Economic Growth and Society*
- *Infant Intervention Programs: Truths and Untruths*
- *Marketing Child Care Programs: Why and How*
- *Competent Caregivers — Competent Children: Training and Education for Child Care Practice*
- *The Feeling Child: Affective Development Reconsidered*
- *Childhood and Chemical Abuse: Prevention and Intervention*
- *Intellectual Giftedness in Young Children: Recognition and Development*
- *The Young Hyperactive Child: Answers to Questions About Diagnosis, Prognosis, and Treatment*

The Young Hyperactive Child: Answers to Questions About Diagnosis, Prognosis, and Treatment

Edited by
Jan Loney, PhD

The Haworth Press
New York • London

The Young Hyperactive Child: Answers to Questions About Diagnosis, Prognosis, and Treatment has also been published as *Journal of Children in Contemporary Society*, Volume 19, Numbers 1/2, Fall/Winter 1986.

© 1987 by The Haworth Press, Inc. All rights reserved. No part of this book may be reproduced or utilized in any form or by any means, electronic or mechanical, including photocopying, microfilm and recording, or by any information storage and retrieval system, without permission in writing from the publisher. Printed in the United States of America.

The Haworth Press, Inc., 12 West 32 Street, New York, NY 10001
EUROSPAN/Haworth, 3 Henrietta Street, London WC2E 8LU England

Library of Congress Cataloging-in-Publication Data

The Young hyperactive child.

"Has also been published as Journal of children in contemporary society, volume 19, numbers 1/2, fall/winter 1986"—T.p. verso.
   Bibliography: p.
   Includes index.
   1. Hyperactive child syndrome. I. Loney, Jan. [DNLM: 1. Attention Deficit Disorder With Hyperactivity—diagnosis. 2. Attention Deficit Disorder With Hyperactivity—therapy. W1 J0584T v.19 no.1/2 / WS 340 Y72]
RJ506.H9Y68   1987   618.92'8589   87-178
ISBN 0-86656-670-8

The Young Hyperactive Child: Answers to Questions About Diagnosis, Prognosis, and Treatment

Journal of Children in Contemporary Society
Volume 19, Numbers 1/2

CONTENTS

About the Editor	ix
Preface	xi
Introduction *Robert L. Sprague, PhD*	1

DIAGNOSIS

How Are DSM-III and DSM-III (R) Used to Make the Diagnosis of Attention Deficit Disorder? *Dennis P. Cantwell, MD*	5
How Is a Parent Rating Scale Used in the Diagnosis of Attention Deficit Disorder? *Thomas M. Achenbach, PhD*	19
How Is a Teacher Rating Scale Used in the Diagnosis of Attention Deficit Disorder? *C. Keith Conners, PhD*	33
How Is a Computerized Attention Test Used in the Diagnosis of Attention Deficit Disorder? *Michael Gordon, PhD*	53
How Is Playroom Behavior Observation Used in the Diagnosis of Attention Deficit Disorder? *Mary Ann Roberts, PhD*	65

PROGNOSIS

What Happens to "Hyperactive" Preschoolers? 75
    Emily K. Szumowski, MS
    Linda J. Ewing, RN, BS
    Susan B. Campbell, PhD

What Are Hyperactive Children Like as Young Adults? 89
    John R. Kramer, PhD

TREATMENT

What Do We Know About the Use and Effects of CNS Stimulants in the Treatment of ADD? 99
    William E. Pelham, Jr., PhD

What Do We Know About the Use and Effects of Behavior Therapies in the Treatment of ADD? 111
    Karen C. Wells, PhD

Cognitive Behavior Therapy for Hyperactive Children: What Do We Know? 123
    Carol K. Whalen, PhD
    Barbara Henker, PhD

What Is the Role of Group Parent Training in the Treatment of ADD Children? 143
    Russell A. Barkley, PhD

What Is the Role of Academic Intervention in the Treatment of Hyperactive Children with Reading Disorders? 153
    Ellis Richardson, PhD
    Samuel Kupietz, PhD
    Steven Maitinsky, MD

SELECTED READINGS 169

Documents and Journal Articles from the ERIC Database 171

Subject Index 173

JOURNAL OF CHILDREN OF CONTEMPORARY SOCIETY Editorial Board

EDITOR

MARY FRANK, *Masters in Education, The Rehabilitation Institute of Pittsburgh*

ASSOCIATE EDITOR

KAREN VANDER VEN, *Professor, School of Social Work, Cathedral of Learning, University of Pittsburgh*

EDITORIAL BOARD

DOROTHY BRYAN, *Developmental Pediatrician, Tertiary Multidisciplinary Service, Children's Hospital of New Jersey*
JOAN COSTELLO, *Associate Professor in the School of Social Service Administration, University of Chicago*
NANCY CURRY, *Professor, School of Social Work, Cathedral of Learning, University of Pittsburgh*
LILLIAN G. KATZ, *Chairman, Department of Elementary and Early Childhood Education and Director of the ERIC Clearinghouse on Elementary and Early Childhood Education, University of Illinois, Urbana-Champaign*
MARGARET McFARLAND, *Associate Professor Emeritus, Department of Psychiatry, School of Medicine, University of Pittsburgh*

INTERNATIONAL EDITORIAL BOARD

FRANK AINSWORTH, *School of Social Work, Phillip Institute of Technology, Bundoora, Victoria, Australia*
HAYDN DAVIES-JONES, *Head of the Further Professional Studies Division, School of Education, University of Newcastle-on-Tyne, Newcastle-on-Tyne, England*
ROY FERGUSON, *Director, School of Child Care, University of Stirling, Scotland, Great Britain*
AVIMA LOMBARD, *Lecturer in Early Childhood Education and Senior Researcher in the NCJW Research Institute for Innovation in Education, School of Education, Hebrew University, Jerusalem, Israel*

ADVISORY BOARD

ELEANOR BARRY, *Supervisory Program Specialist, Pittsburgh Board of Public Education*
WILLIAM HIGNETT, *Director, Louise Child Care Center, Pittsburgh*
WILLIAM ISLER, *Vice President/General Manager, Family Communications, Inc., Pittsburgh*
MARSHA POSTER, *Director, Carnegie-Mellon University Child Care Center, Pittsburgh*
JUDITH RUBIN, ATR, *Western Psychiatric Institute & Clinic, Pittsburgh*
ETHEL M. TITTNICH, *Clinical Assistant Professor, Program of Child Development and Child Care, School of Health Related Professions, University of Pittsburgh*

About the Editor

Jan Loney was born in California and educated at Huntington Beach High School and Stanford University. She obtained a Ph.D. in Child Clinical Psychology from the University of Illinois (Urbana) and subsequently was a post-doctoral fellow at Langley-Porter Neuropsychiatric Institute in San Francisco. During 20 years in the Department of Psychiatry at the University of Iowa, she carried out longitudinal studies of hyperactive children from childhood to adulthood. She is now a Professor in the Department of Psychiatry and Behavioral Science at the State University of New York at Stony Brook, where she is studying childhood hyperactivity and raising race horses.

Preface

The intention of this issue/volume is to pose a series of important questions about hyperactive children and to provide clinicians, educators, and parents with pragmatic and scientifically sound answers derived from the research findings and clinical experience of a group of pre-eminent scientist-practitioners. Over the past 20 years, a massive body of scientific writing has accumulated on the diagnosis, prognosis, and treatment of hyperactive children. As the knowledge base has expanded and as questions concerning childhood hyperactivity have become more sophisticated, there has been a need for detailed and specific summaries of the implications of this enormous body of literature for those whose responsibilities are to treat, educate, and rear these children and to advise their parents and teachers.

The issue/volume is divided into three sections. The first section concerns diagnosis and assessment, the second considers prognosis from pre-school to young adulthood, and the third covers different treatment approaches. Each of the articles/chapters offers an account of the procedures of the particular diagnostic method, prediction strategy, or treatment approach. The general message of the volume is threefold: (1) that comprehensive assessment across informants and situations is required; (2) that individualized prediction of outcomes is both feasible and desirable; and (3) that a long-term combination of treatments tailored to each child's strengths and problems is necessary.

The assessment section begins with a description of diagnosis using the Diagnostic and Statistical Manual of the American Psychiatric Association (DSM-III) and its forthcoming revision (DSM-III-R). Cantwell contrasts the approaches of DSM-III with the contemplated approach of DSM-III-R and points out both strengths and weaknesses in the DSM-III methodology. He then provides an account of collection and integration of the information on which a psychiatric diagnosis is ultimately based. Subsequent articles in this section provide rationales for and practical details on the use of: (1) a parent rating scale, such as the Achenbach Child Behavior Checklist; (2) a teacher rating scale, such as the Conners Teacher Rating

© 1987 by The Haworth Press, Inc. All rights reserved.

Scale; (3) structured observation in a clinic playroom, such as Roberts' SOAPS procedures; and (4) a computerized attention test, such as that developed by Gordon. The methods proposed by Roberts and by Gordon each represent ways of obtaining systematic and relatively objective data on core behavioral symptoms, while the widely-used rating scale approaches described by Conners and by Achenbach offer standardized methods of obtaining the judgments of central adults in the child's natural environment. The different authors advocate the use of multiple measures in the assessment of hyperactive children, partly because more parsimonious approaches have not yet been developed, and partly because childhood hyperactivity is a complex and variable problem in which both objective and impressionistic data are important.

Both articles/chapters in the section on prognosis address questions of crucial importance to treatment and parents as planners. Szumowski, Ewing, and Campbell discuss the problems of differentiating between early signs of a hyperactive behavior disorder, on one hand, versus identifying, on the other hand, the expressions of a transient developmental phase, a problem in family interaction, or a reaction to environmental factors. Kramer discusses the same problem of predicting outcome at another point in the developmental sequence by asking if features at initial referral predict outcomes in young adulthood. It is becoming apparent that the development of a clinically significant disorder and its continuation into adolescence and adulthood is the complex product of interactions between the child's behavioral and cognitive characteristics, the various environmental circumstances in which the child is placed, and the social relationships and interactions that the child experiences.

In the section on treatment, several major intervention strategies are reviewed. Pelham presents a detailed account of current knowledge about treatment with stimulant medication and its role in the management of childhood hyperactivity. Wells gives a similar overview of clinical behavioral therapies and their use in a combined treatment program. Wells also provides an account of the use of single-subject evaluation procedures in monitoring behavioral therapies. Barkley provides a step-by-step description of a training program designed to provide parents with means for dealing with problems in interacting with their oppositional and hyperactive children. Whalen and Hanker give a thoughtful account of the pluses and minuses associated with cognitive approaches to therapy, as well as a set of guidelines for cognitive behavioral therapists. Finally, Richardson, Kupietz, and Maitinsky describe an instructional method for improving reading in hyperactive children. In each of these articles/chapters, the authors emphasize the necessity for tailoring a combination of treatments to the child's individual problems, and they point out the importance of monitoring progress across time and of continuing to provide treatment as long as it is required. It is abundantly clear that no single treatment can be offered as a panacea for all of the problems of all hyperactive children.

As is typical of writing on this topic by different authors or at different points in time, there are differences and inconsistencies in terminology. We have chosen to title the book as one concerning hyperactive children. Childhood hyperactivity is the most generic and descriptive term in common use, and it is ordinarily considered to describe the population of overactive, inattentive, impulsive, and unruly children who often become the focus of adult concern. It avoids the etiological implications of terms such as minimal brain dysfunction, while emphasizing that the recently-derived diagnostic criteria which result in a diagnosis of Attention Deficit Disorder were not necessarily used to diagnose all of the children being discussed. It is obvious that the large heterogeneous group of what are sometimes called externalizing or disruptive children has not yet been subgrouped in a workable and widely-accepted fashion. Differences in terminology across chapters/articles have been allowed to remain. This tolerance of inconsistency is an accurate reflection of the state of the field. The result, however, is that in some cases different terminologies will be applied to very similar groups of children, while, in other cases, identical terminologies will actually apply to quite different groups of children.

As is also typical with a volume/issue of this kind, the editors' acknowledgments extend to many individuals. To begin with, they owe a major debt to each of the individual contributors. In this case, the willingness of a group of nationally-recognized clinical investigators to distill the accumulated knowledge on these topics is most appreciated. The fact that the contributions are pragmatic, balanced, thoughtful, and sound is a tribute to the individual scientific rigor and clinical skills of the contributors themselves. The editor and book-series editor, Mary Frank, owe a major debt of gratitude to Margaret Culkin, whose unflagging attention to detail and cheerful willingness to rise to the occasion did much to keep the issue/volume moving along toward its completion. Finally, the editor, the authors, and the field in general owe a major debt to the individual children and parents in each of the accumulated studies that make up the literature on the complex and challenging topic of childhood hyperactivity. The willingness of these children and their parents to reveal their thoughts and contribute their time in the hope that other families would benefit is something for which we should all be grateful.

Introduction

It is both an honor and privilege and yet a chore to write introductory remarks for a book, particularly when the book covers a broad range of topics with chapters written by numerous authors. Unless one allows the introduction to ramble on for a number of pages, it really is not possible to address each chapter individually and acknowledge appropriately each author's contribution. With those limitations in mind, I plunge into the business of writing this introduction.

This book primarily addresses diagnostic and treatment issues associated with attention deficit disordered children. The editor has solicited numerous authors who represent a good mix and balance of perspectives on diagnosis and treatment. Other dimensions of the problem are covered from a wide perspective also (e.g., the development aspect from preschool to adults is discussed). A wide range of treatment modalities is considered, more than is usualy covered in such a book, including academic performance, particularly reading, computerized assessment, play behavior, to parent training. Many chapters include illustrative case histories which cogently reinforce points made in the chapter and add considerable vitality and interest to what sometimes might be dull academic points.

Several of the themes of the book impress me. Early in the first chapter, it was pointed out that the current diagnostic nomenclature (DSM III) is based upon committee decisions rather than empirically derived information gathered from comprehensive experimental studies. This point is interesting because most of the chapters in the book are heavily empirical, a viewpoint which I certainly share. It must be pointed out, however, that in order to get on with the business of diagnosis as is sorely needed in the clinical area, it is often necessary to start with a concensus opinion of the best experts available due to large gaps in our knowledge base as pointed out by several authors.

The emphasis throughout is on balanced, comprehensive, and empirical data that can be best summarized under the acronym of SCOPE (systematic, comprehensive, objective, practical and efficient), a term previously introduced by the editor (Loney, 1980).

© 1987 by The Haworth Press, Inc. All rights reserved.

Eight chapters are devoted to teacher ratings; parent ratings; the unique domain, at least as it applies to children, of play; and probably the wave of the future, computerized attention measures. A forceful case is made for the necessity of diagnoses to be comprehensive with particular care given to sources of information. Although the point of comprehensiveness in sources of information needs to be made, I must pose the other side of the coin which is just how much of an incremental increase in diagnostic accuracy do we obtain with the additional cost and expense of obtaining a very comprehensive battery of tests, interviews? I must hasten to add, however, that my questions can not be answered now, empirically; rather it is a philosophical question that probably should be kept in mind as more and more extensive batteries of tests, and observations are designed.

More and more it is coming to be realized that attention deficit disorder is a very long-term disorder. Chapters in the book certainly point to this aspect starting with preschool children and continuing through young adulthood. The appropriate central theme of this long-term view of the disorder is a prediction of future status. As discussed in chapter 6, this is a particularly crucial problem with preschool children who in one or two years may present their problems and need special assistance in the primary grades.

Many books are available which focus on one or perhaps two therapeutic modalities. It is certainly refreshing to read about a comprehensive range of therapies in addition to stimulant medication including behavior modification, cognitive behavior therapy, reading remediation, and parent training as it pertains to treating their children. It is stated explicitly several times, and it seems to be the current view, that there is no single therapeutic panacea, but a range of therapeutic treatments are needed with a unique mix adapted for the needs of the individual child. Under this topic, Whalen mentions a number of important considerations in her superbly written chapter. She points out the lack of concern about individual differences when considering treatments and particularly individual differences. She specifically mentions the limitations of her topic, namely cognitive behavior therapy, and it would be well for each of us who write about treatment in this area to keep firmly in mind, and articulate well, the limitations of the treatments we espouse. Whalen also mentions the area of consumer satisfaction which has all but been ignored in the development of treatments for attention deficit disordered children. On the other hand, she mentions expectations about outgrowing the diagnosis which is a point of view that needs serious rethinking because it might be quite unrealistic to expect these children to improve to the point of falling within the normal range of behavioral functioning as has been cogently pointed out by Sleator and Pelham (1986, p. 172).

For any physician, psychologist, mental health worker, or special education teacher/administrator seeking a broad range of information about the problem of hyperactivity now called attention deficit disorder, this book is highly recommended.

Robert L. Sprague, PhD
Professor
Department of Psychology
University of Illinois
Champaign, Illinois

REFERENCES

Loney, J. (1980). Hyperkinesis comes of age: What do we know and where should we go? *American Journal of Orthopsychiatry, 50*, 28-42.

Sleator, E. K., & Pelham, Jr., W. E. (1986). Attention deficit disorder. In D. Cornfeld & B. K. Silverman (Eds.), *Dialogues in pediatric management* (Vol. 1, No. 3). Norwich, CT: Appleton-Century-Crofts.

DIAGNOSIS

How Are DSM-III and DSM-III (R) Used to Make the Diagnosis of Attention Deficit Disorder?

Dennis P. Cantwell, MD

ABSTRACT. Diagnostic issues with the syndrome of Attention Deficit Disorder are reviewed. DSM-III and the proposed DSM-III-R classification schemes are similar in several ways. However, they differ in the specific diagnostic criteria to be used and in subtyping of the disorder. Developmental and gender issues in symptom pattern are discussed. Clinical processes and procedures, diagnostic tools and criteria of the disorder are reviewed.

INTRODUCTION

This article will review the diagnosis of the attention deficit disorder syndrome. Issues to be discussed include: the classification of this disorder, its possible subtypes, the diagnostic criteria specified in DSM-III and DSM-III-R, and the diagnostic process involved in making this diagnosis.

Dennis P. Cantwell, M.D., Joseph Campbell Professor of Child Psychiatry, Neuropsychiatric Institute: Center for the Health Sciences, University of California at Los Angeles, 760 Westwood Plaza, Los Angeles, CA 90024.

© 1987 by The Haworth Press, Inc. All rights reserved.

DSM-III AND DSM-III-R CLASSIFICATION OF THE ATTENTION DEFICIT DISORDER SYNDROME

DSM-III was published in 1980 and offered the first official diagnostic classification system of psychiatric disorders of childhood to specify criteria for the diagnosis of each disorder. It also offered a multiaxial approach to diagnosis in which the clinical psychiatric syndrome was to be coded on Axis I, developmental disorders on Axis II, relevant biological factors on Axis III, relevant psychosocial factors on Axis IV, and adaptive functioning on Axis V.

DSM-III is in the process of being revised, and an initial draft of DSM-III-R has been published. At this point, it is relevant to compare the DSM-III criteria and the DSM-III-R criteria for the diagnosis of attention deficit disorder syndrome.

DIAGNOSTIC CRITERIA

The DSM-III criteria for attention deficit disorder with hyperactivity, attention deficit disorder without hyperactivity, and attention deficit disorder residual type are listed in Table 1. The proposed DSM-III-R criteria for the diagnosis of attention deficit-hyperactivity disorder are somewhat different. Note that in the DSM-III criteria, three subtypes are postulated: with hyperactivity, without hyperactivity, and residual state. The behavioral criteria for the diagnosis of with hyperactivity require that specific symptoms be present in three major areas: inattention, impulsivity, and hyperactivity. Three out of five in the inattention area, three out of six in the impulsivity area, and two out of five in the hyperactivity must be present for the diagnosis of attention deficit disorder with hyperactivity. The criteria for the diagnosis of the without hyperactivity subtype are the same except that the individual does not have criteria C, those of hyperactivity.

The diagnostic criteria for attention deficit disorder residual type include that the individual has had a diagnosis of ADD *with* hyperactivity (not without hyperactivity) in the past, that hyperactivity no longer be present, but that the other signs of the illness be present at the time of the residual diagnosis in adolescence or adult life.

As currently proposed the DSM-III-R criteria are substantially different in a number of ways. There is a list of 15 behavioral symptoms that are thought to characterize the disorder. For the diagnosis to be made, any 8 must be present. (The number 8 is tentative at the

TABLE 1

Diagnostic Criteria for Attention Deficit Disorder with Hyperactivity

The child displays, for his or her mental and chronological age, signs of developmentally inappropriate inattention, impulsivity, and hyperactivity. The signs must be reported by adults in the child's environment, such as parents and teachers. Because the symptoms are typically variable, they may not be observed directly by the clinician. When the reports of teachers and parents conflict, primary consideration should be given to the teacher reports because of greater familiarity with age-appropriate norms. Symptoms typically worsen in situations that require self-application, as in the classroom. Signs of the disorder may be absent when the child is in a new or a one-to-one situation.

The number of symptoms specified is for children between the ages of eight and ten, the peak age range for referral. In younger children, more severe forms of the symptoms and a greater number of symptoms are usually present. The opposite is true of older children.

A. *Inattention.* At least three of the following:
 1. often fails to finish things he or she starts
 2. often doesn't seem to listen
 3. easily distracted
 4. has difficulty concentrating on schoolwork or other tasks requiring sustained attention
 5. has difficulty sticking to a play activity
B. *Impulsivity.* At least three of the following:
 1. often acts before thinking
 2. shifts excessively from one activity to another
 3. has difficulty organizing work (this not being due to cognitive impairment)
 4. needs a lot of supervision
 5. frequently calls out in class
 6. has difficulty awaiting turn in games or group situations
C. *Hyperactivity.* At least two of the following:
 1. runs about or climbs on things excessively
 2. has difficulty sitting still or fidgets excessively
 3. has difficulty staying seated
 4. moves about excessively during sleep
 5. is always "on the go" or acts as if "driven by a motor"
D. Onset before the age of seven.
E. Duration of at least six months.
F. Not due to Schizophrenia, Affective Disorder, or Severe or Profound Mental Retardation.

Diagnostic Criteria for Attention Deficit Disorder Without Hyperactivity

The criteria for this disorder are the same as those for Attention Deficit Disorder with Hyperactivity except that the individual never had signs of hyperactivity (criterion C).

Table 1 (continued)

Diagnostic Criteria for Attention
Deficit Disorder, Residual Type

A. The individual once met the criteria for Attention Deficit Disorder with Hyperactivity. This information may come from the individual or from others, such as family members.
B. Signs of hyperactivity are no longer present, but other signs of the illness have persisted to the present without periods of remission, as evidenced by signs of both attentional deficits and impulsivity (e.g., difficulty organizing work and completing tasks, difficulty concentrating, being easily distracted, making sudden decisions without thought of the consequences).
C. The symptoms of inattention and impulsivity result in some impairment in social or occupational functioning.
D. Not due to Schizophrenia, Affective Disorder, Severe or Profound Mental Retardation, or Schizotypal or Borderline Personality Disorders.

Note. From *Diagnostic and Statistical Manual of Mental Disorders, Third Edition* (pp. 43, 44, 45) by the American Psychiatric Association, 1980. Washington, D.C., American Psychiatric Association. Copyright 1980 by the American Psychiatric Association. Reprinted by permission.

current time, and it is based on a field trial used to determine the sensitivity and specificity of the number of symptoms from the 15 best distinguishing the ADD syndrome from conduct disorder and oppositional disorder.)

Thus, DSM-III and DSM-III-R are similar in that they specify the symptoms, and they specify a certain number for the diagnosis of ADD to be made. They differ in that the ADDH diagnosis requires symptoms in each of three areas: inattention, impulsivity, and hyperactivity. The ADD without hyperactivity DSM-III disorder requires symptoms from the inattention and impulsivity area and does not allow symptoms in the hyperactivity area. The ADD residual state in DSM-III requires current attentional and impulsivity problems and a past history but no current history of hyperactivity.

The DSM-III-R criteria allow any permutation or combination of 8 of the 15 listed symptoms. They do not require that symptoms of inattention, impulsivity, and motor activity be present, and they do not make a distinction between ADDH, ADD without hyperactivity, and ADD residual state.

The latter decision was made on the basis of recent follow-up studies showing that hyperactivity indeed may persist, it is not the symptom which always disappears. Rather, when any 2 out of 3 of the original symptom areas (inattention, impulsivity, or hyperactivity) are present, the syndromal quality of the disorder seems to remain relatively constant in adolescents and young adults.

The criteria are also similar in that there is an age of onset criterion and a duration criterion. Any age of onset, like any specific symptom number, is relatively arbitrary, but the criteria are meant to emphasize that this is an early onset condition with a developmental course. This is not something which begins *de novo* at the age of 10, 11, or 12. The duration criteria emphasize that this is a relatively chronic disorder and is not a transient phenomenon that may be seen in children, for example, with anxiety disorders or affective disorders who may manifest ADD symptoms during the time they are extremely anxious or during the time when they are depressed.

In addition to inclusion criteria, there are also exclusion criteria. That is, when the symptoms are present, the diagnosis is not made if the presence of the excluding conditions such as schizophrenia, affective disorder, or severe or profound mental retardation are present. Any exclusion criteria are relatively arbitrary. If schizophrenia and ADD coexist, it may be difficult to tell what is *due* to schizophrenia and what is a coexistent disorder. The "due to" conventions in DSM-III are being reconsidered in DSM-III-R.

The specificity of the symptom picture in both the DSM-III and the DSM-III-R criteria have certain implications for diagnosis. In the past, terms like hyperactive child syndrome, hyperkinetic reaction of childhood, minimal brain damage, and minimal brain dysfunction were used as almost synonymous terms for this disorder. When behavioral criteria are not specified, however, the same term may be used to describe children with different disorders and different terms may be used to describe children who have the same disorder. Thus, both the DSM-III and DSM-III-R criteria add specificity to the diagnostic picture. If the symptoms are present in the requisite amount, the age of onset is there, the duration is there, and exclusion criteria are not present, then the diagnosis can be more comfortably, probably more reliably, and probably more validly made than when one uses simply a term with no specific diagnostic criteria. This is the value of a DSM-III or a DSM-III-R type classification system.

There are, however, also some drawbacks to both sets of criteria. Both are "made by committee" criteria. They are not criteria that have been developed on an empirical basis by studies of large groups of children to determine the precise symptom pattern that best distinguishes this disorder from children who have no psychiatric disorder and from children who have other types of psychiatric disorder, particularly those with other disruptive behavior disorders such as conduct disorder or oppositional disorder. Although there was an attempt with a field trial for the DSM-III-R criteria to pick out the number of symptoms best distinguishing DSM-III-R, ADD-hyperactivity syndrome from conduct disorder and oppositional disorder as diagnosed by DSM-III-R criteria. The field trial was limited, had some methodological problems, and the specific symptoms picked for each of the conditions—ADD, conduct, and oppositional were "made by committee" symptoms.

Other symptoms that may be important in this disorder are not described as part of the essential features in either the DSM-III or DSM-III-R criteria. For example, many investigators feel that children with the attention deficit disorder syndrome have important difficulties with peer relationships and a "lack of social savoir-faire" in dealing with other children. The children may recognize and may feel bad about the fact that other children do not like them, but they have a hard time understanding what it is that they do to contribute to the fact that other children do not get along with them.

There is also a developmental issue. Children are developing organisms. It is likely that the same phenomenologic picture is not necessarily going to be present with the same symptom pattern in preschool children, early grade school children, late grade school children, early adolescents, and late adolescents. Some mention is made of this in DSM-III in that it is stated in the text that in younger children, more symptoms are usually present and that in older children fewer symptoms may be present. Systematic studies have not been done, however, to determine exactly which symptoms are more likely to be present in preschoolers, grade school, and adolescent children. This is a problem for all phenomenologically based diagnostic symptoms in that the child is a developing organism, and the number of symptoms or the expression of the symptoms may change with age.

Gender issues may also be involved. It may be that girls, for example, are more likely to have certain symptoms than boys, although this has not been systematically studied to any great degree. These developmental issues and possible gender issues add to the relative arbitrariness of a specific number of symptoms needing to be made for the diagnosis to be present. For example, no one would argue with the fact that a child with 7 of the 15 DSM-III-R specific symptoms who has had these symptoms consistently since toddler age and in whom the symptoms are causing a great deal of functioning impairment should not have the diagnosis made and proper intervention instituted.

DIAGNOSTIC EVALUATION PROCESS

Specifying a certain number of specific symptoms for the diagnosis can aid in the reliability of a diagnosis. This has important implicaions for both clinical practice and research. Much more is involved, however, in making a diagnosis than simply having a specific set of diagnostic criteria. In addition to the diagnostic classification system which specifies particular criteria for the diagnosis and possibly for subtypes of the diagnosis, the other issues involved in the diagnostic process are: what questions one is trying to answer as a clinician, what tools one has to collect data to answer those questions, and how does one assimilate the data collected to making a particular diagnosis? A complete discussion of all of these issues is beyond the scope of this article. Each of these is treated more fully in references listed at the end of the article.

Questions To Be Considered

When a child is presented for evaluation with a particular problem, there are a series of questions that the clinician asks. One of these, of course, is: Is a particular disorder present? That essentially means does the child meet DSM-III or DSM-III-R criteria for the diagnosis of a specific disorder such as the attention deficit-hyperactivity disorder? Does the child's disorder meet the criteria for other disorders as well? Many ADD children will in addition meet criteria for developmental learning disorders, developmental communication disorders, conduct disorders, and oppositional disorders. The case study described at the end of this article illustrates some of these issues.

In addition to the question of which specific disorder the child presents, there are other questions that are important to answer in the diagnostic evaluation process. Some of these questions include the following: What are the likely etiologic roots of the disorder in this particular individual child? These roots may be biological in nature, they may be genetic, they may be familial-nongenetic, they may be psychosocial, they may be intrapsychic, or any combination of these. These need to be explored, and these are above and beyond whether or not the child has the disorder. What forces are in the child himself and in the child's family and greater psychosocial environment that are helping to maintain the problem? Conversely, what forces are helping to facilitate the child's normal development? What are the individual strengths of the child? What are his or her competencies? What are the strengths of the family, the school, and the greater psychosocial environment? These latter three questions are important for treatment planning. The last three questions are probably the most important from the standpoint of clinical intervention. One of these is the likely natural history of this child's disorder. We know a substantial amount now about the untreated natural history of the ADD syndrome, and we know something about the predictive factors that are associated with a good or poor prognosis.

Finally, the clinician needs to ascertain whether intervention is necessary in this case and what types of intervention are likely to be effective and likely to be safe. Intervention is necessary when the child's disorder is causing a significant degree of functional impairment and is impeding normal developmental growth. The likely natural history of the child's disorder is also important in determining the urgency of the intervention.

To determine what interventions are likely to be effective and likely to be safe, the clinician needs to turn to the literature which is generally organized around diagnostic syndromes. To the extent that there are specific diagnostic criteria to make a diagnosis, the literature is likely to be more solid with regard to treatment histories. If the diagnostic criteria are used, at least one can be assured that children with the same clinical picture are being treated identically in diverse clinical and research settings. If only a diagnostic term is used with no specified diagnostic criteria, differences in natural history and differences in response to various types of therapeutic intervention may be a function of the fact that the children who are being treated have different disorders rather than that the treatment is or is not effective.

Diagnostic Tools

The last 10 years or so have seen the development of diagnostic tools to aid the clinician in making specific diagnoses in child psychiatry. Many of these were developed for the diagnosis of the ADDH syndrome in psychopharmacologic studies. They can be divided into: interviews with parents, interviews with the children, behavior rating scales, the physical exam, the neurological exam, and various laboratory studies. A complete discussion of all of these instruments is beyond the scope of this article, but each of them are discussed more fully in the references provided at the end of the article.

The making of a specific diagnosis, that is, determining whether or not specific criteria are fulfilled, will generally be done with the interview with the parents, the interview with the child, and with the use of behavior rating scales. Some laboratory measures may be useful in quantifying attentional difficulties or motor activity, but these generally will not be used in the ordinary clinical setting.

The interview with the parent can conveniently be divided up into six major areas: the referral source, the chief complaint, the symptom inventory of recent behavior, assessment of the child's temperament and personality characteristics, past medical history and developmental history, and family history. The section on symptom inventory and recent behavior will generally be the area where specific diagnostic criteria are assessed. There are standardized diagnostic interviews for parents and also for children, many of which are parent and child versions of the same interview such as the DICA, the DISC, and the Kiddie-SADS, which are useful for quantifying the symptom picture. It is important, however, for the clinician to flesh out the symptom picture and ask for a more detailed description from the parents of the symptoms that have been listed during the interview. It is mandatory that a recent example of each of the symptoms be provided, as well as the frequency of symptom occurrence, the severity of each symptom, and the content of its occurrence. If there are certain factors that seem to make certain symptoms better, they should always be noted as well as what methods the parents have attempted to use in dealing with a particular symptom.

There are many reasons for fleshing out this symptom picture, but the most important is that what is "hyperactive," "inattentive," "impulsive" to one parent may be quite different to another parent. It is imperative during the interview with the parent that the clinician differentiate what actually happened and what symptoms were present from how the parents felt about what happened and what feelings were generated by a particular set of behaviors. Retrospective recall is important to keep in mind. Parents generally will remember better whether a symptom has occured than when it has occured, and they are usually better able to delineate major symptoms than minor symptoms.

The interview with the child, which may take place over more than one setting, will generally be used to assess many of the same areas. In order to make the diagnosis of the ADD syndrome, the symptoms should be assessed from the child as well. They may be assessed from direct observations which are spontaneous observations during the interview setting or elicited observations. One can do certain things to assess attention, impulsivity, motor activity, etc. in the interview setting. Also, verbal material will be obtained during the interview with the child. These may be spontaneous, and they may be elicited by direct questions, particularly if one is using a standardized semistructured or structured interview. The child's developmental stage, his intellectual level, his language, and his conceptual abilities must be assessed in order to determine whether or not certain concepts are understood, and what weight to place on the child's report of symptoms. In general, children older than 6 years of age can give accurate accounts of behavior that is observable to the parent. Systematic studies have shown about an 80% agreement on the questions of the DICA, with no tendency to increase agreement with age. Children, however, are better sources of certain types of information such as internal experiences of mood, suicidal ideation, and delusional material. Parents may be better sources of information of observable behavior such as inattentiveness, impulsive behavior, and hyperactivity. Thus, there may be a disagreement between the parental interview, the data gathered from the parent, and the data gathered from the child in the interview setting, and this will have to be rectified.

Behavior Rating Scales

The third data collection source useful for making the diagnosis of the attention deficit disorder syndrome are various behavior rating scales. There are parent rating scales, teacher rating scales, and multi-informant rating scales which have been useful. The original Conners parent rating scale, the revised version, and the abbreviated form for parents have all been found to be useful. The Achenbach child behavior checklist is more comprehensive but has not yet been used as much in diagnosis and monitoring of the ADD syndrome. There is an original, a revised, and an abbreviated Conners teacher rating scale and an Iowa Conners teacher rating scale developed by Loney and her colleagues which has been useful in distinguishing children with the attention deficit disorder syndrome from conduct and oppositional disordered children.

The rating scales are useful for quantification of the child symptoms in various settings by various raters. As a general approach, it is important to get information from various raters in various settings, if for no other reason than that children's behavior tends to differ from setting to setting, especially children who have the ADD syndrome.

In addition, various raters may have different access to different types of information. Teachers may be a better source of rating attentional behavior in the "real world" than parents simply because they have a number of children who are doing the same tasks, a built-in control group, in addition to the fact that the classroom setting may put more stress on the attentional mechanism.

The physical and neurological examination are not generally useful in making a specific behavioral diagnosis but are useful in assessing biologic factors that would be diagnosed on Axis III and which may be useful in subdividing children with this syndrome.

In addition, laboratory measures are not generally useful in deciding whether or not specific symptoms are present. They may, however, be useful as a way of quantifying sustained attention, selective attention, cognitive impulsivity, motor activity, and other such constructs.

There must be an assimilation and integration of all of the data collected from different sources to make the diagnosis. In the ordinary clinical setting, the clinician assesses the information obtained from multiple data sources, and integrates them into a diagnostic statement and a diagnostic formulation. An advance in the field would be a way of formalizing the integration of the data from various sources and possibly looking at conceptual goals, relating the assessment process to the diagnostic classification system with its various disorders. To date this has not been done.

SUMMARY AND DISCUSSION

This article has briefly reviewed some of the issues involved in the diagnostic evaluation of children with the ADD syndrome with particular reference to DSM-III and DSM-III-R criteria. Specific criteria are likely to lead to increased reliability and ideally to increased validity. The diagnostic process itself, the tools used in that process, and the assimilation of the data collected are also important issues. The latter two aspects have been under researched.

CASE HISTORY

J.M. was a 3-1/2 year old boy when referred for evaluation. Nursery school teachers noted that J was unable to sit on the rug at reading time as long as other children. He often seemed to be "not listening" or "day dreaming." At times teachers thought he might be hard of hearing and at other times he seemed to hear normally.

His motor development was slightly below age level and he seemed to be "always on the go." His articulation was immature and his expressive vocabulary seemed to be below that of his classmates.

On the playground, J was impulsive and tended to wander around from one group to the next, not staying at any one activity for any length of time.

History from the parents as well as information from parent and teacher rating scales indicated a long-term pattern of problems with attention, concentration, cognitive and behavioral impulsivity, motor coordination, activity level, and speech and language development. Intelligence testing revealed an IQ in the normal range but a verbal IQ 15 points lower than performance IQ. Speech and language evaluation revealed delays in articulation and expressive language, with a lesser delay in language comprehension. Gross motor coordination was mildly delayed.

J was placed in a pre-school class for communication disordered children, and the parents entered into a Parent Training Program. One year later J was re-evaluated. His articulation and language scores were now in the normal range and his verbal IQ had increased. However, his classroom and home behavior remained relatively unchanged. At this point J was begun on a trial of psychostimulant medication.

He responded dramatically with increased sustained and selective attention, decreased impulsivity and activity. After a second year in the communication pre-school, J was able to attend regular kindergarten and progressed well.

REFERENCES

American Psychiatric Association. (1980). *Diagnostic and statistical manual of mental disorders* (3rd ed.). Washington, DC: Author.

American Psychiatric Association (1985). *DSM-III-R in development.* Washington, DC: Author.

Cantwell, D. P. (1975). *The hyperactive child—diagnosis, management, and current research.* New York: Spectrum Publications.

Cantwell, D. P. (1980). The diagnostic process and diagnostic classification in child psychiatry—DSM-III. *Journal of the American Academy of Child Psychiatry, 19*, 345-355.

Cantwell, D. P. (in press). Clinical child psychopathology—Diagnostic assessment, diagnostic process, and diagnostic classification: DSM-III studies. In M. Rutter, H. Tuma, & I. Lann (Eds.), *Diagnosis and classification of child psychiatric disorders.* New York: Guilford Press.

Conners, S. K. & Barkley, R. (1985). Rating scales and checklists for child psychopathology. *Psychopharmacology Bulletin, 21*(4), 809-815.

Hunt, R. (in press). Attention deficit disorder—diagnosis, assessment, and treatment. In C. J. Kestenbaum & D. T. Williams (Eds.), *Clinical assessment of children and adolescents: A biopsychosocial approach.* New York: New York University Press.

Klorman, R. (1985). Some thoughts on the diagnosis of ADDH in adolescents. *Psychopharmacology Bulletin, 21*(4), 913-914.

National Institute of Mental Health. (1985). Rating scales, and assessment instruments for use in pediatric psychopharmacology research (Special issue). *Psychopharmacology Bulletin, 21*(4), U.S. Department of Health and Human Services Publication No. (ADM) 86-173.

Orvaschel, H. (1985). Psychiatric interviews suitable for use in research with children and adolescents. *Psychopharmacology Bulletin, 21*(4), 737-746.

Reatig, N. (1985). Bibliography on rating and assessment instruments for attention deficit disorder (ADD). *Psychopharmacology Bulletin, 21*(4), 929-932.

Ross, D. M., & Ross, S. A. (1982). *Hyperactivity: Current issues, research, and theory* (2nd ed.). New York: Wiley & Sons.

Rutter, M. (Ed.). (1983). *Developmental neuropsychiatry.* New York: Guilford Press.

Whalen, C. K. (1983). Hyperactivity, learning problems, and attention deficit disorders. In T.H. Ollendick & M. Hersen (Eds.), *Handbook of child psychopathology* (pp. 159-171). New York: Plenum Press.

How Is a Parent Rating Scale Used in the Diagnosis of Attention Deficit Disorder?

Thomas M. Achenbach, PhD

ABSTRACT. Parents are important sources of data about children's problems, and their views usually determine what will be done about the problems. In combination with teachers' reports, parents' reports of hyperactivity are especially important, because hyperactive behavior is often not evident in the clinical setting. Parent rating scales are designed to provide a standardized format for obtaining a differentiated picture of parents' perceptions of their child. As illustrated with the Child Behavior Checklist, parent ratings can be scored on profiles that compare a parent's report of problems and competencies with those by parents of normative groups of agemates. Parent rating scales can be used whenever information is obtained from parents about their child. They are especially useful as part of initial intake evaluations and for reassessment to evaluate treatment effects and follow-up status. A multiaxial approach to assessment integrates parent ratings with data from sources such as teachers, direct observations, medical exams, psychometrics, and clinical interviews.

Before addressing the "how" of using a parent rating scale, we should be clear about *why* we use a parent rating scale and *what* we are trying to assess with it. The term "diagnosis" implies the identification of a specific disorder by a clinician. Yet, the decision to seek help for a child and most of the data on the child's problem behavior come from nonclinicians who interact with the child in everyday environments. The most important of these nonclinicians are usually the child's parents and teachers. Even the DSM-III diagnostic criteria for attention deficit disorders state that the "signs must be reported by adults in the child's environment, such as

---

Thomas M. Achenbach is Professor and Director, Center for Children, Youth and Families, Department of Psychiatry, University of Vermont, Burlington, VT 05401.

© 1987 by The Haworth Press, Inc. All rights reserved.

parents and teachers . . . . because . . . . they may not be observed directly by the clinician" (American Psychiatric Association, 1980, p. 43).

Reports by parents and teachers are both important for the different perspectives they bring to bear on children's functioning in two important developmental arenas, the home and the school. What parents report about their child often differs from what teachers report, and both may differ from what the child reports and what the clinician observes. (Achenbach, McConaughy & Howell, 1987, summarize findings from over 100 studies on the levels of agreement between different informants.) Because a child's behavior often differs from one situation to another, and different informants perceive the child differently, disparities between reports by different informants do not necessarily mean that one is wrong and the other is right. Instead, because no one person can observe a child across all situations and all interaction partners, it is important to obtain a clear picture of how the child is seen by each of those most involved with him or her. Parents can usually report on the widest variety of their child's behavior over the longest periods of time. Parents' perceptions are also especially important in determining whether a child will receive help and, if so, what kind and for how long.

But how do we know whether what is reported really indicates deviance? If we merely ask parents to tell us about their child, their replies will reflect their verbal fluency, educational level, previous contacts with mental health services, and general family dynamics, as well as the child's behavior. If the parents are especially concerned about a particular behavior or have already "diagnosed" their child with a label such as "hyperactive," it may be hard to obtain a differentiated picture of other aspects of the child's behavior. It may also be hard to determine how much their child actually deviates from other children and how their perceptions might differ from those of other people who know the child.

THE CHILD BEHAVIOR CHECKLIST

To optimize the use of parents' reports, rating scales obtain data from parents in a standardized format that facilitates comparison with normal children and with reports by other informants, such as teachers, clinicians, and the children themselves. Although there are numerous parent rating scales, we will focus on the Child Behavior Checklist (CBCL) as our main example (Achenbach & Edelbrock, 1983, present details of the standardization and use of the CBCL). Unlike scales designed to assess a single type of disorder, such as attention deficits or depression, the CBCL includes a wide variety of problems that are easily reported by most parents. Examples include *Gets in many fights, Fears going to school*, and *Unhappy, sad, or depressed*, as well as problems indicative of hyperactivity, such as *Can't sit still, restless, or hyperactive*, and attention deficits, such as *Can't concentrate, can't pay attention for long*. For each problem item, parents circle a *0* if the item is *not true* of their child, *1* if it is *somewhat or sometimes true*, and *2* if it is *very true or often true*. Parents are asked to base their judgments on the previous six months, but other rating periods can be used if desired.

Because children's adaptive competencies are as important as their problems in determining their need for help, the CBCL also has items for assessing the child's involvement in sports, other activities, organizations, jobs and chores, friendships, relations with significant others, and school functioning. The CBCL requires only a sixth grade reading level and can be filled out by most parents in about 15 minutes. If a parent cannot read, the CBCL can be read aloud by an interviewer who records the parents' responses. For parents who do not know English, there are translations in 23 other languages.

The Child Behavior Profile for Scoring the CBCL

The CBCL is designed to obtain standardized descriptions of children's behavior as seen by their parents. To relate the parents' ratings to particular syndromes, however, and to compare the child's standing with norms for children of the same age and sex, the CBCL is scored on the Child Behavior Profile. The Profile consists of scales for scoring competence in terms of activities, social involvement, and school functioning, plus either eight or nine problem scales, depending on the age and sex of the child. The problem scales reflect syndromes of problems found to occur together in CBCLs filled out by the parents of 2,300 children who had been referred for mental health services. To reflect age and sex differences in the patterning and prevalence of problem behaviors, versions of the Profile are standardized separately for each sex at ages 4 to 5, 6 to 11, and 12 to 16 years. Norms for the syndrome scales are based on parents' CBCL ratings of 1,300 randomly selected children who had not been referred for mental health services (Achenbach & Edelbrock, 1981). Table 1 lists the syndrome scales of the Profile for children of each age and sex. Several of the empirically-derived syndromes are quite similar to DSM syndromes such as attention deficit, conduct, depressive, and anxiety disorders. DSM diagnoses made from structured psychiatric interviews have been found to correlate well with the corresponding

Table 1

SYNDROMES FOUND THROUGH FACTOR ANALYSIS
OF THE CHILD BEHAVIOR CHECKLIST

Group	Internalizing Syndromes	Mixed Syndromes	Externalizing Syndromes
Boys aged 4-5	Social Withdrawal Depressed Immature Somatic Complaints	Sex Problems	Delinquent Aggressive Schizoid
Boys aged 6-11	Schizoid or Anxious Depressed Uncommunicative Obsessive-Compulsive Somatic Complaints	Social Withdrawal	Delinquent Aggressive Hyperactive
Boys aged 12-16	Somatic Complaints Schizoid Uncommunicative Immature Obsessive-Compulsive	Hostile Withdrawal	Hyperactive Aggressive Delinquent
Girls aged 4-5	Somatic Complaints Depressed Schizoid or Anxious Social Withdrawal	Obese	Hyperactive Sex Problems Aggressive
Girls aged 6-11	Depressed Social Withdrawal Somatic Complaints Schizoid-Obsessive		Cruel Aggressive Delinquent Sex Problems Hyperactive
Girls aged 12-16	Anxious-Obsessive Somatic Complaints Schizoid Depressed Withdrawal	Immature Hyperactive	Cruel Aggressive Delinquent

Child Behavior Profile syndromes (Edelbrock, 1984). Despite their similarities, there are also the following differences between the DSM approach to diagnosis and the CBCL approach:

1. The DSM definitions were developed through committee negotiations, whereas the CBCL syndromes were derived from statistical analyses of the problems actually reported for large samples of children.
2. The DSM does not specify particular assessment procedures for determining whether a child qualifies for a diagnosis such as attention deficit disorder, whereas parents' ratings on the CBCL directly provide syndrome scores.
3. The DSM requires disorders to be diagnosed as present versus absent, whereas the CBCL syndromes are scored in a quantitative fashion, enabling us to view a child in terms of the *degree to which* he or she manifests a syndrome such as hyperactivity.
4. The DSM provides no basis for comparing a child with normal peers, whereas the CBCL syndrome scales provide percentiles and standard scores (T scores) for comparison with normal peers.

HOW TO USE A PARENT RATING SCALE

A parent rating scale can be used whenever information is obtainable from parents or parent surrogates about a child's behavior. Having parents fill out a rating scale such as the CBCL is a natural part of the intake procedure for mental health services. If parents fill out the CBCL before the first interview, the clinician can use the completed form and scored profile as a takeoff point for interviewing. Rather than wasting valuable interview time to obtain an initial description of the child's behavior, the clinician can obtain from the CBCL a differentiated, standardized picture of the parents' perceptions of the child's behavior.

By examining the profile scored from the CBCL, the clinician can identify the areas in which the parents' reports indicate more problems or fewer competencies than are reported by most parents of normal children. It is important to consider and compare what is reported in all the areas relevant to a child's age and sex, rather than focusing exclusively on one area such as hyperactivity. Otherwise, the parents' concerns about one area can mask problems in other areas that may be more important.

It is also helpful to have the child's mother and father fill out separate CBCLs, when possible. By comparing profiles scored from both parents' CBCLs, similarities and differences between the parents' perceptions of the child can be identified. Although mothers' scores do not usually differ much from fathers' scores, large differences between the profiles obtained from a child's mother and father can be clinically informative. As an example, Figure 1 shows profiles scores from CBCLs completed by the mother (solid line) and father (broken line) of an 8-year-old third-grader named Joey. Joey's teacher had urged his parents to have him evaluated for hyperactivity, because of his failure to pay attention, to stay seated, and to learn in school. Joey's parents were each asked to fill out the CBCL as part of the intake procedure at their local child guidance clinic.

The profiles scored from Joey's mother's CBCL and father's CBCL agree in showing somewhat more problems of the hyperactive syndrome than reported by parents of 98% of normal (i.e., nonreferred) 6-11-year-old boys. (The 98th percentile is used as a cutoff to distinguish between the clinical and normal range, as shown by the broken line printed across the profile.) Besides the problems of hyperactivity, however, both parents' profiles show scores above the 98th percentile on the Depressed and Delinquent scales. Joey's mother's profile also shows more deviance on the Aggressive scale than any other scale, whereas his father's profile shows a score on the Aggressive scale that borders between the clinical and normal range.

In asking the parents about the differences between their reports of aggressive behavior, the clinician learned that several specific behaviors of the Aggressive scale, including *Argues a lot*, *Screams a lot*, and *Threatens people* occurred in the mother's presence but not in the father's presence. Other than this difference, the parents' profiles generally agreed in showing approximately as much deviance on the Depressed and Delinquent scales as on the Hyperactive scale. The social competence portion of their profiles also agreed in indicating that activities were within the normal range, but that social relations and school functioning were below the 2nd percentile for normal 6-11-year-old boys. As Figure 2 shows, the border between the normal and clinical range on the social competence scales is at the 2nd percentile, with the clinical range comprising scores that are lower than those obtained by 98% of children in the normative samples.

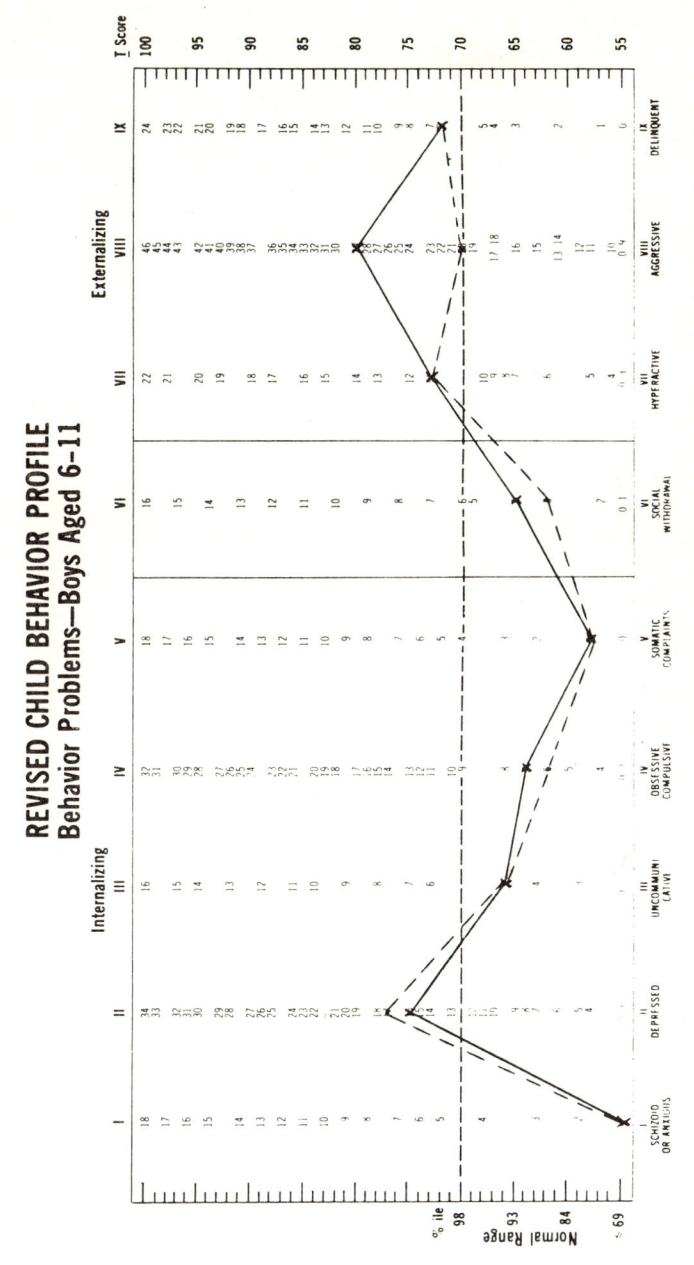

FIGURE 1. Behavior problem profiles scored from CBCLs completed by the mother (solid line) and father (broken line) of an 8-year-old third-grader named Joey.

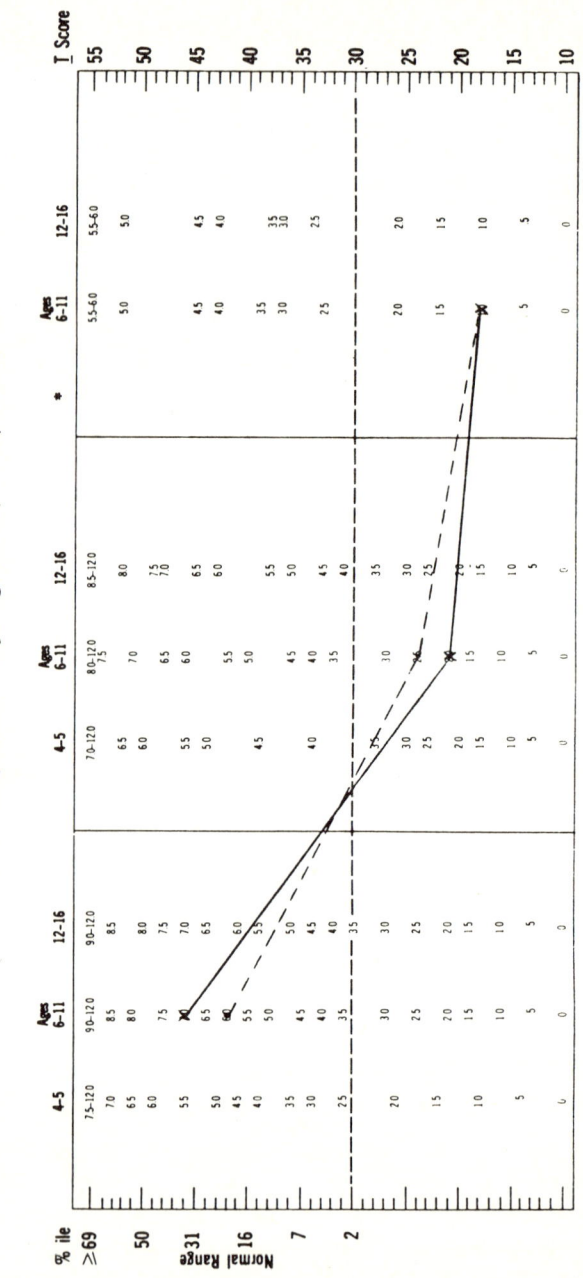

FIGURE 2. Social competence profiles scored from CBCLs completed by the mother (solid line) and father (broken line) of 8-year-old Joey.

Because teachers' perspectives are also important, especially in cases like Joey's where school problems are a concern, Joey's teacher was asked to complete the Teacher's Report Form (TRF) of the Child Behavior Checklist. The TRF is scored on the Teacher's Report Version of the Child Behavior Profile. The teacher profile consists of scales for problems and adaptive functioning derived from ratings by teachers of 1,700 pupils referred for special school and mental health services. Norms are based on TRFs completed by teachers of 1,100 randomly selected nonreferred pupils (Achenbach & Edelbrock, 1986, present details of the standardization and use of the TRF).

Figure 3 shows the behavior problem portion of the teacher profile scored from the TRF completed by Joey's teacher. Because the problem scales were derived from ratings by teachers, they reflect correlations between behaviors as seen by the teachers. Some of the scales differ, therefore, from those derived from parents' ratings. The Unpopular scale on the teacher profile, for example, has no counterpart on the parent profile, because teachers are more apt to observe the problem patterns comprising that scale than parents are. Furthermore, two separate syndromes—designated as Inattentive and Nervous-Overactive—were found in teacher ratings, whereas items from these syndromes were combined in the Hyperactive syndrome found in the parent ratings. This distinction between problems of attention and overactivity doubtless emerged in teacher ratings because teachers have more occasion to judge children's inattentiveness to tasks requiring concentration. Teacher profile patterns showing deviance mainly on the Inattentive scale versus those showing deviance on the Overactive and other scales have been found to discriminate between children independently diagnosed as attention deficit without hyperactivity and those diagnosed as attention deficit with hyperactivity (Edelbrock, Costello & Kessler, 1984). The teacher profile also discriminated between these two groups and children who were not diagnosed as having attention deficits.

In Joey's case, the profiles indicated that overactivity and inattention were part of a picture that included at least as much deviance in depressive and delinquent behavior, as seen by both his parents, unpopularity as seen by his teacher, and greater deviance in aggressive behavior as seen by his mother and teacher. Although the problems of hyperactivity and inattention reported by his parents and teachers would qualify Joey for a diagnosis of attention deficit disorder with

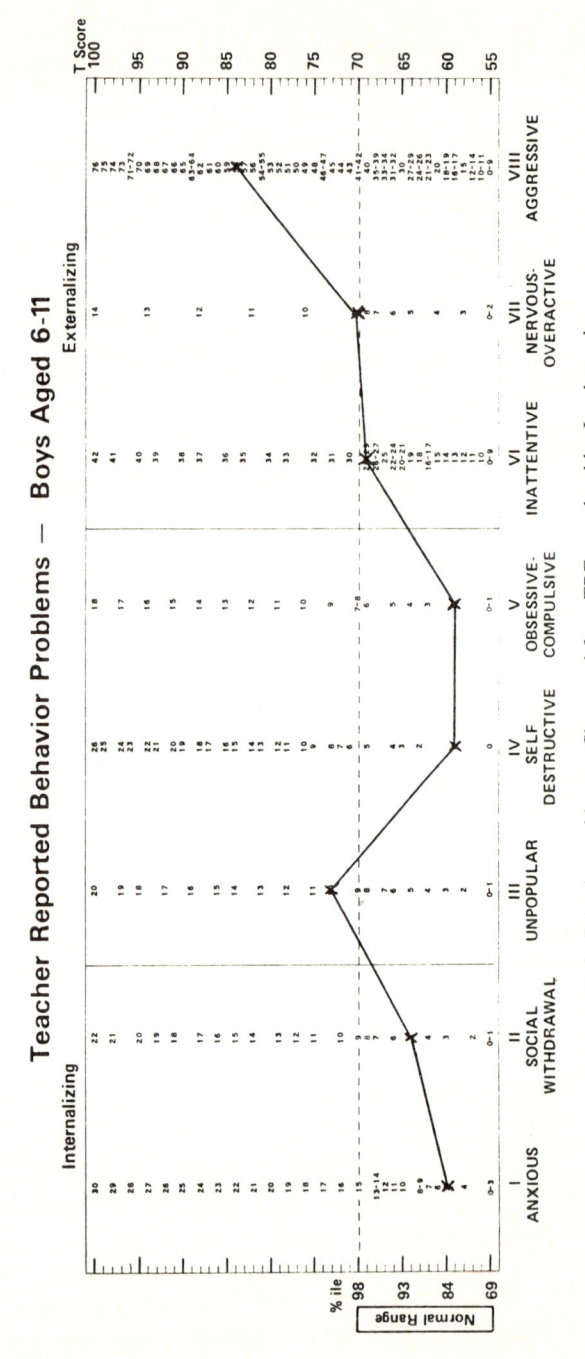

FIGURE 3. Behavior problem profile scored from TRF completed by Joey's teacher.

hyperactivity, the profiles show that it would be wrong to stereotype Joey as a "hyperactive child." Instead, any intervention plan should take account of the depressive and delinquent behavior shown at home, his unpopularity with peers evident at school, and the high levels of aggression in the presence of his mother and teacher, but less so with his father.

Reassessments of Children

Although a parent rating scale such as the CBCL forms a natural part of initial evaluation procedures, it also provides a baseline against which to compare a child's functioning at later points. After an intervention is begun, for example, it is important to reassess the child periodically to determine whether he or she is improving in the targeted areas and not getting worse in other areas. In Joey's case, this could be done by having his parents complete the CBCLs and his teacher complete the TRF after the intervention has time to work and new behavior has time to emerge and stabilize. A 3- to 6-month period after the initial ratings is usually appropriate.

The profiles scored from the reassessment CBCLs and TRF can be compared with the original profiles to determine how much change occurs on all scales. After the therapeutic goals have been achieved, it is also desirable to obtain follow-up ratings 6 months to 18 months after the initial ratings. The follow-up profiles can then be compared with the earlier ones to determine whether progress is maintained. Because the profiles show percentiles and standard scores for comparison with normative samples of agemates, the user can determine not only whether there are improvements, but also whether what is reported about the child's behavior reaches the normal range. This is important, because significant improvements alone may not be sufficient to bring a child's behavior into the normal range.

QUESTIONS TO BE ASKED OF A PARENT RATING SCALE

Because parents are the main source of data about most children's behavior and parents' judgments are often decisive, it is essential to optimize the use of parents' reports. This does not mean that parents' reports outweigh other types of data, but only that they should be obtained in a standardized form for integration with data from other sources such as teachers, direct observations, medical exams, psychometrics, and clinical interviews. (For a multiaxial approach to assessment incorporating these different sources of data, see Achenbach, 1985.) The types of questions to be asked of a comprehensive parent rating scale and scoring profile include the following:

1. When provided with an extensive list of problems and competencies, which ones do parents report as characterizing their child?
2. Do the parents' ratings raise questions that should be explored clinically to obtain a fuller description, to discuss parents' interpretations of reasons for particular behaviors, or to determine reasons for discrepancies between a mother's and father's report?
3. When scored on a standardized profile, what areas of problems or competencies indicate deviance from normative groups of children?
4. What are the major similarities and differences between the parents' reports and those obtained in a standardized fashion from others, such as teachers, observers, clinicians, and the children themselves?
5. When scored on a standardized profile, what does the child's overall pattern look like? Are problems restricted to a single area such as hyperactivity, or is help required in multiple areas?
6. Which areas of problems or competencies should we focus on first? If a child is below the normal range with respect to competencies, as well as having significant problems, this might argue for interventions to strengthen competencies. For a child whose competencies are in the normal range, on the other hand, interventions should probably focus on the problem areas.
7. Based on comparisons of parents' ratings with other sources of data, is there evidence that the parents' perceptions of the child require changing more than the child's behavior does?
8. Do post-treatment ratings indicate that the child has reached the normal range or that areas of deviance remain despite possible improvements?
9. Do follow-up ratings show that progress is maintained?

Although this overview of parent rating scales may be too brief to equip the reader to use them without further preparation, more information on the scales discussed here can be obtained by writing to the author at the Department of Psychiatry, University of Vermont, 1 South Prospect Street, Burlington, VT 05401.

REFERENCES

Achenbach, T.M. (1985). *Assessment and taxonomy of child and adolescent psychopathology.* Beverly Hills, CA: Sage.

Achenbach, T.M., & Edelbrock, C.S. (1981). Behavioral problems and competencies reported by parents of normal and disturbed children aged four to sixteen. *Monographs of the Society for Research in Child Development, 46* (Serial No. 188).

Achenbach, T.M., Edelbrock, C. (1983). *Manual for the child behavior checklist and revised child behavior profile.* Burlington, VT: University of Vermont Department of Psychiatry.

Achenbach, T.M., & Edelbrock, C. (1987). *Manual for the teacher's report form and teacher version of the child behavior profile.* Burlington, VT: University of Vermont Department of Psychiatry.

Achenbach, T.M., McConaughy, S.H., & Howell, C.T. (1986). *Child/adolescent behavioral and emotional problems: Implications of cross informant correlations for situational specificity.* Manuscript submitted for publication.

American Psychiatric Association (1980). *Diagnostic and Statistical Manual of Mental Disorders* (3rd ed.). Washington, DC: Author.

Edelbrock, C. (1984, October). *Relations between the NIMH Diagnostic Interview Schedule for Children (DISC) and the Child Behavior Checklist and Profile.* Paper presented at the meeting of the American Academy of Child Psychiatry, Toronto.

Edelbrock, C., Costello, A.J., & Kessler, M.D. (1984). Empirical corroboration of attention deficit disorder. *Journal of the American Academy of Child Psychiatry, 23,* 285-290.

How Is a Teacher Rating Scale Used in the Diagnosis of Attention Deficit Disorder?

C. Keith Conners, PhD

ABSTRACT. Teachers can provide important and useful information to assist in the diagnostic process for Attention Deficit Disorder (ADD). The ideal conditions for teacher information are seldom met, and rating scales are a useful adjunct when personal familiarity with the teacher is precluded by practical considerations. Rater biases, positive and negative halo effects, practice effects and other problems associated with rating scales are outweighed by the ease of use, low cost, and reasonable reliability and validity of teacher rating scales. Problems of over- and under-diagnosis can be avoided by reliance upon a comprehensive assessment of which teacher information is an important part.

Clinicians have long understood that information from teachers is vital for making correct diagnoses of school-age children. The importance of teacher information for the diagnosis of Attention Deficit Disorder (ADD) was formalized by DSM-III (The Diagnostic and Statistical Manual of the American Psychiatric Association, 1980):

> The signs must be reported by adults in the child's environment, such as parents and teachers . . . When the reports of teachers and parents conflict, primary consideration should be given to the teacher reports because of greater familiarity with age-appropriate norms. Symptoms typically worsen in situations that require self-application, as in the classroom. (p. 43)

C. Keith Conners is Professor of Psychiatry and Behavioral Sciences, Department of Psychiatry, George Washington University School of Medicine, and Children's Hospital National Medical Center. Address reprint requests to the author at Children's Hospital National Medical Center, 111 Michigan Ave., N.W., Washington, DC 20010.

© 1987 by The Haworth Press, Inc. All rights reserved.

Presumably the ideal method of data gathering from teachers would include the following:

1. establish a friendly and trusting relationship with the teacher;
2. evaluate the teacher's objectivity as an informant, general understanding of psychopathology, and knowledge of the particular child;
3. evaluate the structure and process of teaching in his or her classroom;
4. carefully interview the teacher regarding the child's strengths and weaknesses;
5. obtain a sufficiently long baseline of the teacher's judgments to insure representativeness and stability of the child's behavior pattern;
6. interview other teachers from previous years or from other parts of the curriculum; and
7. place the findings in the context of what is known about the cultural and demographic aspects of this school system.

Such an approach is certainly possible where the clinician is, say, a school psychologist or physician who regularly works within a public school setting as a consultant; or where the school is a special school within an inpatient or residential facility. For most clinicians, however, such an ideal procedure is unrealistic. Both clinicians and teachers are generally too busy and costs are too high to make these procedures feasible under ordinary circumstances.

In addition, there are drawbacks to such a highly individualized approach. Standardized comparisons across a broader context become difficult. Comparisons with other children of the same age are limited to those within a single classroom or school. Research data and norms will be unavailable. Many children in poorly-run classrooms will be considered deviant, or contrarily, in settings where many severely deviant children attend school, few may be seen as special problems since the "norm" is one of deviance.

Given these problems, teacher rating scales would appear to provide a simple and economic method of obtaining relevant information as part of the diagnostic process. They allow a common vocabulary for describing academic, social and emotional behaviors in the classroom. They are generally economical of teacher and clinician time, have norms based on large samples of normal children of different ages and social backgrounds, and have been widely used in validity comparisons among different diagnostic groups. The latter is particularly true for ADD and other "externalizing" disorders.

PROBLEMS WITH TEACHER RATING SCALES

Some authors are troubled by teacher rating scales as a source of diagnostic information because they believe that the scales mistakenly give the *appearance* of objective data by assigning numerical scores to judgments which reflect only "subjective" impressions of teachers (Carey & McDevitt, 1980). This objection is actually a philosophical position which contrasts supposedly "objective" data with "subjective" impressions. In this context, subjectivity is used in a pejorative sense to mean either unreliability or lack of validity.

Unfortunately, there is no "teacher-meter" that can provide better information than ratings. It is precisely the teacher's *judgment* that is required. Judgments are the stuff of which all diagnoses are eventually made, even in medical disciplines where laboratory-based information is more readily available as a source of data. The same questions of reliability and validity apply whether the instrument is the human observer making global judgments, or quantitative frequency and time-sampling counts; whether the construct being studied is mechanically measured (as by an actometer), or an impression conveyed by a verbal statement. In many studies, teacher ratings have been shown to be both more reliable and valid than other presumably more "objective" methods.

More tenable objections are that global rating scales require sufficient knowledge of the child being rated, a criterion not always met, and they are subject to halo and rater bias effects (Beitchman & Raman, 1979). Both of these objections are well-taken, but there are numerous reasons why *self-report* rating scales, as recommended by these authors, are an inadequate substitute. The teacher's knowledge of the child, and the clinician's knowledge of the teacher are *both* presupposed when using teacher ratings in a diagnostic decision, and it is up to the clinician to evaluate the source of this and any other information used in diagnostic formulations.

WHAT DO TEACHER RATING SCALES MEASURE?

Numerous studies have been conducted on the reliability and validity of teacher rating scales. (For a comprehensive review, see Barkley, 1986. For a review of rating scales for ADD see Reatig,

1985). These scales have been compared with measures of activity in structured and unstructured settings, with actometers, and with a variety of other instruments, including direct classroom observation. While mixed, the results of these studies generally support the utility and cost-effectiveness of teacher rating instruments for measuring behaviors generally associated with ADD/H, especially impulsive verbal and motor behaviors, and on-task classroom activity.

With regard to the specific factors of our teacher rating scale (Conners, 1969) that bear on the question of the diagnosis of ADD (with and without hyperactivity), good evidence exists that more "objective" measures such as direct observation of deviant classroom behaviors are substantially correlated with the hyperactivity and inattention factors (Copeland & Weissbrod, 1978, 1980; Firestone & Douglas, 1975; Aaron, 1979; Abikoff, Gittelman-Klein & Klein, 1977, 1980; Aman, 1979; Badian, 1977).

Several findings from these studies deserve mention. Abikoff et al. (1977, 1980) developed a reliable frequency and time-sampling observation code for hyperactivity. The standard deviations of the mean scores for the comparison and hyperactive children on the Conners Teacher Questionnaire indicated no overlap between children referred for treatment of hyperactivity and a normal control group. There *was* overlap, however, in the direct observation code data, and the authors suggest that diagnosis should be based on teacher ratings and *not* on the direct behavior code information. Considering that the cost of the rating data is substantially less than the labor-intensive process of training and acquiring direct observational data, it would seem axiomatic that teacher ratings are to be preferred in the usual clinical diagnostic process.

Secondly, although the concept of "Attention Deficit Without Hyperactivity" is currently under scrutiny for possible elimination from DSM-III (it has proven very difficult to diagnose with any reliability); several studies indicate that teacher rating factors of inattentiveness and hyperactivity do discriminate children who have attentional problems without hyperactivity (Aaron, 1979; Badian, 1977; Aman, 1979).

Finally, although teacher rating scales have been most useful in identifying and studying hyperactive/ADD children, they are really sensitive to the entire "externalizing" dimension of behavior (Achenbach & Edelbrock, 1978), and consequently have been found in a number of studies to include children who also have conduct disorders. In addition, because items relating to internalizing behavior (anxiety, mood) are *under*-represented, the scale will be less sensitive to the presence of anxiety than to ADD. Consequently, some children who appear to have ADD may also have undetected anxiety disorders, or children with anxiety disorders may be missed.

STEPS IN USING A TEACHER RATING SCALE

1. *Qualitative Analysis*

Figure 1 presents the most recent version of our 39-item teacher rating scale.

FIGURE 1. A Teacher Rating Scale

CHILDREN'S HOSPITAL NATIONAL MEDICAL CENTER
PRELIMINARY SCHOOL REPORT*

OFFICE USE
Patient No.
Study No.

Name of Child _____ Date _____
School Attended _____ Grade _____
School Address _____
  Number and Street   City   State
Name of Principal _____

I. How long have you known this child? _____ In your own words describe briefly this child's main problem. _____

II. STANDARDIZED TEST RESULTS

A. Intelligence Tests

Name of Test	Date	C.A.	M.A.	I.Q.

B. Most Recent Achievement Tests

Subject	Grade When Tested	Achievement Grade Level
Reading		
Spelling		
Arithmetic		

Figure 1, continued

III. ACHIEVEMENT IN SCHOOL SUBJECTS

A. List subjects into the appropriate category.

Very Good	Average	Barely Passing	Failing

B. Check special placement or help this child has received.

( ) Ungraded ( ) Sight-Saving ( ) Special Class ( ) Remedial Reading ( ) Speech Correction

( ) Tutoring, specify subjects _____

( ) Other, specify _____

*Report design by C. Keith Conners, Ph. D.

Form 7

IV. Listed below are descriptive terms of behavior. Place a check mark in the column which best describes this child. ANSWER ALL ITEMS.

Observation	Degree of Activity			
	Not at all	Just a little	Pretty much	Very much
CLASSROOM BEHAVIOR				
1. Constantly fidgeting				
2. Hums and makes other odd noises				
3. Demands must be met immediately—easily frustrated				
4. Coordination poor				
5. Restless or overactive				
6. Excitable, impulsive				
7. Inattentive, easily distracted				
8. Fails to finish things he starts—short attention span				
9. Overly sensitive				
10. Overly serious or sad				
11. Daydreams				
12. Sullen or sulky				
13. Cries often and easily				
14. Disturbs other children				
15. Quarrelsome				
16. Mood changes quickly and drastically				
17. Acts "smart"				
18. Destructive				
19. Steals				
20. Lies				
21. Temper outbursts, explosive and unpredictable behavior				

Figure 1, continued

GROUP PARTICIPATION				
22. Isolates himself from other children				
23. Appears to be unaccepted by group				
24. Appears to be easily led				
25. No sense of fair play				
26. Appears to lack leadership				
27. Does not get along with opposite sex				
28. Does not get along with same sex				
29. Teases other children or interferes with their activities				

ATTITUDE TOWARD AUTHORITY				
30. Submissive				
31. Defiant				
32. Impudent				
33. Shy				
34. Fearful				
35. Excessive demands for teacher's attention				
36. Stubborn				
37. Overly anxious to please				
38. Uncooperative				
39. Attendance problem				

V. FAMILY OF CHILD

   A. Do other children in the family who attend your school, present any problems?
      If YES, please explain. _____

   B. Please add any information concerning this child's home or family relationships which might have bearing on his attitudes and behavior, and include any suggestions for improvement of his behavior and adjustment. (Use reverse side if more space is required.)

Figure 1, continued

|Signature | Title | Date Signed|

The first step is to examine the overall pattern of response by the teacher: does s/he check mostly toward the "Not At All" side or more in the direction of "Very Much?" This will give an overall impression of how severe the teacher thinks the child's problems are. Generally, most normal children will have "Not at All" or "Just A Little" checked; in other words, it is quite normal for a child to be a little inattentive or restless. It is unusual for normal children to have behaviors which are checked "Very Much," or who have a substantial number of "Pretty Much" items. One must keep in mind that some teachers tend to see very little pathology in *any* children, while others see it everywhere.

Now one can focus on those items which are more specifically relevant to the major symptoms required for a DSM-III diagnosis (impulsivity, inattention, and hyperactivity). Some of the items required as minor symptoms in DSM-III are specifically included in the teacher scale, such as "often fails to finish things he or she starts," while others are approximately the same (e.g., "easily distracted" in DSM-III is close to "7. Inattentive, easily distracted"). In this way, one may get a rough idea whether there are moderate or severe symptoms to support the required pattern for a diagnosis in DSM-III. Unfortunately, since most teacher scales were composed prior to DSM-III, they only approximate the formal symptom requirements.

One of the problems with a qualitative approach is that individual items might be strongly endorsed, but the overall pattern may not suggest real pathology. It is also difficult to arrive at a quantitative estimate so that one child can be compared with another. For this reason, it is better to combine certain items which tend to cluster together statistically, which increases the reliability of assessing a "dimension" or cluster of behaviors, and then compare these "factors" with norms obtained on large samples.

2. The Normative Approach

Each item is scored 0, 1, 2, or 3, and the items belonging to a factor are summed to get a factor score. Table 1 presents the scoring key for the six factors of our teacher scale. Tables 2 through 7 present normative tables in the form of T-scores (mean = 50, standard deviation = 10). After summing the items belonging to a factor, each raw score for a given factor is entered into the appropriate age and sex column to obtain the T-score. In general, a T-score of 70 or greater is considered clinically significant; this score represents a score of two or more standard deviations from the mean. The norms for these factors are based upon the study by Trites, Blouin and Laprade, 1982 on a sample of 9583 public school children in Canada (excluding French-speaking children).

As noted above, in making a diagnosis of ADD, the teacher report is only one of several sources of information to be considered. When a child has a high Conduct Disorder score as well as a high Hyperactivity factor score, other information will be required to determine: (a) whether the conduct disturbance is secondary to a true hyperactivity problem; (b) whether the child has both a primary conduct disorder *and* hyperactivity; or (c) whether a teacher's negative halo effect is operating for a troublesome child. The daydreaming-inattentive factor has been useful in identifying children whose learning problem is associated (and perhaps caused by) inattentiveness without hyperactivity.

One common source of error in teacher ratings of hyperactivity is a *positive halo effect*, in which a very bright or engaging child is characterized as "exuberant" or "vigorous," either because the teacher wishes to avoid labelling the child, is against any suggestion that medication might be indicated, or is unfamiliar with the other

TABLE 1

FACTORS OF THE 39-ITEM TEACHER RATING SCALE

FACTOR TITLE	ITEM	# ITEMS
I. Hyperactivity	1, 2, 3, 4, 5, 6, 7, 8, 11, 14, 15, 17, 24, 29, 32, 35, 38	17
II. Conduct Disorder	15, 16, 17, 18, 19, 20, 21, 25, 29, 31, 32, 36, 38	13
III. Emotional Indulgent	3, 9, 10, 12, 13, 16, 21, 36	8
IV. Anxious Passive	24, 26, 30, 33, 34, 37	6
V. Asocial	22, 23, 25, 27, 28	5
VI. Daydream	8, 11, 22, 39	4
*** Hyperactivity Index	1, 3, 5, 6, 7, 8, 13, 14, 16, 21	10

TABLE 2

Conners Teacher Rating Scale—39 Item

Hyperactivity

Score	\multicolumn{9}{c}{Males Ages}				\multicolumn{9}{c}{Females Ages}			
	4	5	6	7	8	9	10	11	12	4	5	6	7	8	9	10	11	12
0	40	40	39	39	39	39	39	40	38	41	42	42	42	41	42	42	43	41
1	41	41	40	40	40	40	40	41	39	43	43	43	43	43	43	43	44	42
2	42	42	41	41	41	41	41	42	40	44	44	44	44	44	44	45	45	43
3	43	43	42	42	42	42	42	43	41	45	45	45	45	45	45	46	46	44
4	44	44	43	43	43	43	43	44	42	46	46	46	46	46	46	47	48	45
5	45	45	44	44	44	44	44	45	43	47	47	47	47	47	47	48	49	46
6	46	46	45	45	45	45	45	46	44	48	48	48	48	48	48	49	50	47
7	47	47	46	46	46	46	46	47	45	49	49	49	49	49	49	50	51	48
8	48	48	47	47	47	47	47	48	46	50	50	50	50	50	50	51	52	49
9	49	49	48	48	48	48	48	49	46	51	51	51	51	51	51	52	53	50
10	50	50	49	49	49	49	49	50	47	52	52	52	52	52	52	53	54	50
11	51	51	50	50	50	50	50	51	48	53	53	53	53	53	53	54	55	51
12	52	52	51	51	51	51	51	52	49	54	54	54	54	54	54	55	56	52
13	53	53	52	52	52	52	52	53	50	55	55	55	55	55	55	56	57	53
14	54	54	53	53	53	53	53	54	51	56	56	56	56	56	56	57	58	54
15	55	55	54	54	54	54	54	55	52	57	57	57	57	57	57	58	59	55
16	56	56	55	55	55	55	55	56	53	58	58	58	58	58	58	59	60	56
17	57	57	56	56	56	56	56	57	54	59	59	59	59	59	59	60	61	57
18	58	58	57	57	57	57	57	58	55	60	60	60	60	60	60	61	62	58
19	59	59	58	58	58	58	58	59	56	61	61	61	61	61	61	62	63	59
20	60	60	59	59	59	59	59	60	57	62	62	62	62	62	62	63	64	60
21	61	61	60	60	60	60	60	61	58	63	63	63	63	63	63	64	65	61
22	62	62	61	61	61	61	61	62	59	64	64	64	64	64	64	65	66	62
23	63	63	62	62	62	62	62	63	60	65	65	65	65	65	65	66	67	63
24	64	64	63	63	63	63	63	64	61	66	66	66	66	66	66	67	68	64
25	65	65	64	64	64	64	64	65	62	67	67	67	67	67	67	68	69	65
26	66	66	65	65	65	65	65	66	63	68	68	68	68	68	68	69	70	66
27	67	67	66	66	66	66	66	67	64	69	69	69	69	69	69	70	71	67
28	68	68	67	67	67	67	67	68	65	70	70	70	70	70	70	71	72	68
29	69	69	68	68	68	68	68	69	66	71	71	71	71	71	71	72	73	69
30	70	70	69	69	69	69	69	70	67	72	72	72	72	72	72	73	74	70
31	71	71	70	70	70	70	70	71	68	73	73	73	73	73	73	74	75	71
32	72	72	71	71	71	71	71	72	69	74	74	74	74	74	74	75	76	72
33	73	73	72	72	72	72	72	73	70	75	75	75	75	75	75	76	77	73
34	74	74	73	73	73	73	73	74	71	76	76	76	76	76	76	77	78	74
35	75	75	74	74	74	74	74	75	72	77	77	77	77	77	77	78	79	75
36	76	76	75	75	75	75	75	76	73	78	78	78	78	78	78	79	80	76
37	77	77	76	76	76	76	76	77	74	79	79	79	79	79	79	80	81	77
38	78	78	77	77	77	77	77	78	75	80	80	80	80	80	80	81	82	78
39	79	79	78	78	78	78	78	79	76	81	81	81	81	81	81	82	83	79
40	80	80	79	79	79	79	79	80	77	82	82	82	82	82	82	83	84	80
41	81	81	80	80	80	80	80	81	78	83	83	83	83	83	83	84	85	81
42	82	82	81	81	81	81	81	82	79	84	84	84	84	84	84	85	86	82
43	83	83	82	82	82	82	82	83	80	85	85	85	85	85	85	86	87	83
44	84	84	83	83	83	83	83	84	81	86	86	86	86	86	86	87	88	84
45	85	85	84	84	84	84	84	85	82	87	87	87	87	87	87	88	89	85
46	86	86	85	85	85	85	85	86	83	88	88	88	88	88	88	89	90	86
47	87	87	86	86	86	86	86	87	84	89	89	89	89	89	89	90	91	87
48	88	88	87	87	87	87	87	88	85	90	90	90	90	90	90	91	92	88
49	89	89	88	88	88	88	88	89	86	91	91	91	91	91	91	92	93	89
50	90	90	89	88	88	88	89	89	87	92	92	92	92	92	92	93	94	90
51	91	91	89	89	89	89	90	90	88	93	93	93	93	93	93	94	95	91

43

TABLE 3

Conners Teacher Rating Scale—39 Item

Conduct Disorder

Score	M4	M5	M6	M7	M8	M9	M10	M11	M12	F4	F5	F6	F7	F8	F9	F10	F11	F12
0	43	44	44	43	43	43	43	43	42	44	45	45	45	44	45	45	45	44
1	45	45	47	45	45	45	45	45	43	46	47	47	48	47	47	47	47	46
2	46	47	48	46	46	46	46	46	44	47	49	49	50	49	49	50	49	47
3	48	49	50	48	48	48	48	48	45	48	50	51	52	50	51	52	51	49
4	49	50	50	50	50	49	49	49	47	50	52	53	54	53	52	54	53	50
5	51	52	51	51	51	51	51	51	48	52	54	55	55	55	54	56	55	51
6	52	54	53	53	53	52	53	52	49	54	56	57	57	57	56	58	56	53
7	54	55	54	55	55	54	54	54	50	56	58	59	59	58	58	60	58	54
8	55	57	56	56	56	55	56	55	51	58	59	61	61	60	60	62	60	55
9	57	59	57	58	58	57	57	57	52	60	61	63	64	62	62	64	62	57
10	58	60	59	59	59	58	59	58	53	62	63	65	66	64	64	66	64	58
11	60	62	60	61	61	60	60	59	54	64	66	67	68	66	66	69	66	60
12	62	63	62	63	63	61	62	61	55	66	68	69	70	68	68	71	68	61
13	63	65	63	64	65	63	63	63	56	67	70	71	73	71	70	73	71	63
14	65	66	65	66	66	64	64	64	57	69	72	73	75	73	72	75	73	64
15	66	68	66	67	68	66	66	66	58	71	75	75	77	75	74	77	75	66
16	68	69	68	69	69	67	67	67	59	73	77	77	79	77	76	79	77	68
17	69	71	69	71	71	69	69	69	60	75	79	79	82	79	78	81	79	69
18	71	72	71	72	73	70	70	70	61	77	81	81	84	81	80	83	81	71
19	72	74	72	74	75	72	72	72	63	79	82	83	86	84	82	85	83	72
20	74	76	74	76	76	73	73	73	64	81	84	85	88	86	85	88	85	74
21	76	77	76	77	78	74	75	75	65	83	86	87	89	88	87	90	87	76
22	77	79	77	79	79	76	76	76	66	85	88	89	91	90	89	92	88	77
23	79	81	78	80	81	78	78	77	67	87	89	91	93	92	91	94	90	78
24	80	82	80	82	83	79	79	79	68	89	91	93	95	95	93	96	92	80
25	82	84	81	84	85	81	81	80	70	90	93	95	98	97	95	98	94	81
26	83	86	83	85	86	82	82	82	71	92	95	97	100	99	97	100	96	82
27	85	87	84	87	88	84	84	83	72	94	98	99	102	101	99	102	98	83
28	86	89	86	88	90	85	85	85	73	96	100	101	105	103	101	104	99	85
29	88	90	87	90	91	87	87	86	74	98	102	103	107	105	103	107	101	86
30	89	92	89	92	93	88	89	88	76	100	103	105	109	108	105	109	103	87
31	91	94	90	93	95	89	90	89	77	102	105	107	111	110	107	111	105	88
32	93	95	92	95	96	91	92	91	78	104	107	109	114	112	109	113	107	89
33	94	97	93	97	98	92	93	92	79	106	109	111	116	114	111	115	109	91
34	96	98	95	98	100	94	95	94	80	108	112	113	118	116	113	117	110	92
35	97	101	96	100	101	95	97	95	81	110	114	115	121	119	114	119	112	93
36	99	102	98	101	103	97	98	97	82	112	116	117	123	121	116	121	114	94
37	100	104	99	103	105	98	100	98	83	114	119	119	125	123	118	123	115	95
38	102	106	100	105	106	100	101	100	85	116	121	121	130	125	120	126	116	96
39	103	107	102	106	108	103	104	101	87	119	123	123	134	129	120	128	118	97

TABLE 4

Conners Teacher Rating Scale--39 Item
Emot. Indulgent

	Males												Females								
Ages:	4	5	6	7	8	9	10	11	12	4	5	6	7	8	9	10	11	12
Score																		
0	41	43	42	43	42	43	42	43	41	42	43	43	43	43	43	43	43	42
1	44	45	45	45	45	45	45	45	43	45	45	45	46	46	45	46	46	44
2	47	48	47	48	47	48	47	48	45	47	48	48	49	48	48	49	50	46
3	49	50	50	50	50	50	50	50	47	49	50	51	53	51	51	52	53	49
4	52	52	52	53	53	52	52	53	49	52	53	54	56	54	53	55	56	51
5	54	55	55	55	55	55	55	55	52	54	55	57	59	57	56	59	59	53
6	57	57	58	58	58	57	57	58	54	56	58	59	62	60	58	62	62	55
7	59	60	60	61	60	59	60	60	56	59	60	62	65	63	61	65	65	57
8	62	62	63	63	63	61	62	63	58	61	63	65	68	66	64	68	68	59
9	65	64	65	66	66	64	65	65	60	64	65	68	71	69	66	71	71	61
10	67	67	68	68	68	66	67	68	62	66	68	71	74	72	69	74	74	63
11	70	69	70	71	71	68	70	70	64	68	70	73	77	75	72	77	77	65
12	72	72	73	73	74	71	72	73	66	71	73	76	80	77	74	80	80	67
13	75	74	76	76	76	73	75	75	68	73	75	79	83	80	77	83	83	69
14	77	76	78	78	79	75	77	78	70	76	78	82	86	83	79	86	86	71
15	80	79	81	81	81	78	80	80	72	78	80	84	89	86	82	89	90	73
16	83	81	83	83	84	80	82	83	74	80	83	87	92	89	85	92	93	76
17	85	84	86	86	87	82	85	85	76	83	85	90	95	92	87	95	96	78
18	88	86	88	88	89	85	88	88	78	85	88	93	98	95	90	98	99	80
19	90	88	91	91	91	87	90	90	81	87	91	96	101	98	93	101	102	82
20	93	91	94	93	94	89	93	92	83	90	93	98	104	101	95	104	105	84
21	96	93	96	96	97	92	95	95	85	92	96	101	107	104	98	107	108	86
22	98	96	99	98	99	94	98	97	87	95	98	104	110	106	100	110	111	88
23	101	98	101	101	102	96	100	100	89	97	101	107	114	109	103	113	114	90
24	103	100	104	103	105	98	103	102	91	99	103	109	117	112	106	116	117	92

45

TABLE 5

Conners Teacher Rating Scale—39 Item

Anxious-Passive

	Males									Females								
Ages:	4	5	6	7	8	9	10	11	12		4	5	6	7	8	9	10	11	12
Score
0	39	39	38	39	39	38	37	39	37		38	39	38	40	39	38	37	39	38
1	42	42	41	42	42	41	41	42	40		41	42	41	43	42	41	40	42	40
2	45	45	43	45	45	44	44	45	43		44	45	44	46	45	43	43	46	43
3	48	48	46	48	48	47	47	49	46		47	47	47	49	48	46	46	49	45
4	48	51	49	51	51	50	50	52	49		49	50	49	51	51	49	49	52	48
5	53	54	52	54	54	53	54	55	53		52	53	52	54	54	52	52	55	50
6	56	56	54	57	57	56	57	59	56		55	55	55	57	57	55	55	58	53
7	59	59	57	60	61	59	60	62	59		57	58	57	60	60	58	58	61	55
8	62	62	60	63	64	62	63	65	62		60	61	60	63	63	60	61	64	58
9	65	65	62	66	67	65	66	69	65		63	64	63	66	66	63	64	67	61
10	67	68	65	69	70	68	70	72	68		66	66	66	69	69	66	67	70	63
11	70	71	68	72	73	70	73	75	71		68	69	68	72	72	69	71	73	66
12	73	74	70	75	76	73	76	79	74		71	72	71	75	75	72	74	76	68
13	76	77	73	78	79	76	79	82	77		74	74	74	78	78	75	77	79	71
14	79	80	76	81	82	79	82	85	80		76	77	77	81	81	77	80	82	73
15	81	83	79	84	85	82	86	88	83		79	80	79	84	84	80	83	85	76
16	84	85	81	87	88	85	89	92	86		82	82	82	87	87	83	86	88	78
17	87	88	84	90	91	88	92	95	89		85	85	85	90	90	86	89	91	81
18	90	91	87	93	94	91	95	98	92		87	88	87	93	93	89	92	94	84

TABLE 6

Conners Teacher Rating Scale—39 Item
Asocial

	Males									Females								
Ages:	4	5	6	7	8	9	10	11	12	4	5	6	7	8	9	10	11	12
Score																		
0	43	44	44	44	43	43	43	44	42	44	44	45	45	44	44	44	45	38
1	47	49	48	49	47	48	47	48	45	48	50	50	50	48	49	49	49	43
2	50	53	53	54	51	52	51	51	49	52	55	54	55	52	53	53	53	49
3	54	58	57	59	55	56	55	55	52	56	60	59	60	56	58	57	58	54
4	58	62	61	63	59	60	59	59	55	60	65	63	66	59	62	62	62	60
5	62	67	65	68	63	64	63	63	59	64	70	68	71	63	67	66	66	65
6	65	71	69	73	67	68	67	67	62	68	75	72	76	67	72	71	71	71
7	69	75	73	78	70	72	71	71	65	72	81	77	81	71	76	75	75	76
8	73	80	77	82	74	76	76	75	68	77	86	81	86	75	81	80	79	82
9	77	84	81	87	78	80	80	79	72	81	91	86	91	78	86	84	84	87
10	81	89	86	92	82	84	84	83	75	85	96	91	97	82	90	89	88	93
11	84	93	90	97	86	88	88	87	78	89	101	95	102	86	95	93	92	98
12	88	98	94	102	90	93	92	91	82	93	106	100	107	90	100	98	97	104
13	92	102	98	106	94	97	96	95	85	97	111	105	112	94	104	102	101	109
14	96	107	102	111	98	101	100	99	88	101	116	109	117	97	109	106	105	115
15	99	111	106	116	101	105	104	103	91	105	122	114	122	101	113	111	109	120

47

TABLE 7

Conners Teacher Rating Scale--39 Item

Daydream

	Males									Females								
Ages:	4	5	6	7	8	9	10	11	12	4	5	6	7	8	9	10	11	12
Score																		
0	41	41	40	41	41	40	41	41	39	43	43	43	43	42	42	42	43	41
1	47	47	45	46	46	45	46	46	43	48	49	48	49	47	48	48	48	45
2	53	52	49	51	51	50	51	51	47	54	54	52	55	53	55	54	54	49
3	58	58	54	56	56	55	56	56	51	60	60	57	61	58	61	60	60	53
4	64	63	59	61	60	60	61	61	56	66	66	62	67	64	68	66	65	58
5	70	69	63	66	65	64	66	66	60	72	72	67	73	69	74	71	71	62
6	76	74	68	71	70	69	71	71	64	78	77	72	79	75	80	77	77	66
7	82	80	72	76	75	74	76	75	68	84	83	77	85	80	87	83	83	70
8	88	85	77	81	80	79	81	80	72	90	89	82	91	86	93	89	88	74
9	94	91	81	87	85	84	86	85	76	96	94	87	97	91	99	95	94	78
10	100	96	86	92	90	89	91	90	80	102	100	92	102	97	106	101	100	82
11	105	102	91	97	95	93	96	95	85	108	106	97	108	102	112	106	105	86
12	111	107	95	102	100	98	101	100	89	114	112	102	114	108	119	112	111	90

48

consequences of this behavior pattern, such as the destructive effects upon peer relationships and family functioning. Once again, it is important that information be cross-checked with other sources. A dilemma arises, however, when data from parent and teacher conflict. There may be a true "situational" hyperactivity, a pattern of behavior which only emerges, say, in the school setting but not the home setting, or one or other of the observers may be denying or misinterpreting the behavior. There is no simple solution, and inevitably one relies upon the entire context of birth, developmental, social, and medical history; examinations of cognitive function; peer reports; interviews with family and child; and if necessary, home and school observation visits.

Although some teachers will over-endorse externalizing behaviors, thus increasing the hyperactivity and conduct disorder scores, it is likely that they will under-recognize passive, withdrawn, depressed and anxious children. Interviews of parents and child are required to adequately assess these problem areas. In any case, *while teacher ratings are indispensable in arriving at a diagnosis of ADD, they are by no means sufficient for making any diagnoses by themselves.*

In most such scales there is a "practice" effect, in which scores are somewhat lower after the first administration. For this reason, as with the Parent Scales, it is advisable to have at least two, and preferably more, baseline measures before making any diagnostic or treatment decisions. This is particularly important if a single child is being followed and is observed before and after an intervention.

3. Using the "Hyperactivity Index"

Ten of the items from the 39-item scale have been composed as a brief scale which is useful for a number of purposes. These items actually represent items from the factor analyses which show the highest loadings (strongest association) with each of the factors. Thus, the scale is not really a hyperactivity index, but a *psychopathology index*. These are in effect the most robust and strongly weighted items from the entire scale. The scale is shown in Table 8.

Figure 2 presents the age and sex norms for this index, again derived from the study by Trites et al. (1982).

The Index, or Abbreviated Teacher Questionnaire (ATQ), is

useful for monitoring a child's behavior over time. Teachers will often resist the lengthier scale after the first time or two, but will usually be cooperative for the task of regularly filling out a much shorter scale. Moreover, the scale is quite sensitive to drug treatment effects. Caution is encouraged when using the scale in diagnostic decisions since it clearly encompasses several distinct aspects of psychopathology, being particularly likely to confound Conduct Disorder and ADD.

TABLE 8

CONNERS RATING SCALE

HYPERACTIVITY INDEX OR THE ABBREVIATED SYMPTOM QUESTIONNAIRE (ASQ)

TABLE OF T-SCORES BY AGE AND SEX

TEACHER

		MALE					FEMALE			
AGE(YRS)		3-5	6-11	9-11	12-14	15-17	3-5	6-8	9-11	12-14	15-17
TOTAL SCORE	0	42	40	40	40	41	39	42	42	43	44
	1	43	42	41	42	43	40	44	44	47	46
	2	44	44	43	44	45	42	46	46	51	47
	3	45	45	44	47	48	43	49	48	55	49
	4	46	47	46	49	50	45	51	50	59	51
	5	47	49	47	51	52	46	53	53	63	52
	6	48	50	49	54	54	48	55	55	68	54
	7	49	52	50	56	56	49	58	57	72	55
	8	50	54	52	58	59	51	60	59	76	57
	9	51	55	54	61	61	52	62	61	79	59
	10	52	57	55	63	63	54	64	63	84	60
	11	53	59	57	65	65	55	66	65	88	62
	12	54	60	58	68	68	57	69	67	93	64
	13	55	62	60	70	70	58	71	69	97	65
	14	56	63	61	72	72	60	73	71	101	67
	15	57	65	63	75	74	61	75	73	105	68
	16	58	67	64	77	76	63	78	75	109	70
	17	59	68	66	79	79	64	80	78	113	72
	18	60	70	67	82	81	66	82	80	117	73
	19	61	72	69	84	83	67	84	82	122	75
	20	62	73	70	86	85	69	86	84	126	76
	21	63	75	72	89	88	70	89	86	130	79
	22	64	77	74	91	90	72	91	88	134	80
	23	66	78	75	93	92	73	93	90	138	81
	24	67	80	77	96	94	75	95	92	143	83
	25	68	81	78	98	96	76	98	94	147	85
	26	69	83	80	100	99	78	100	96	151	86
	27	70	85	81	103	101	79	102	98	155	88
	28	71	86	83	105	103	81	104	100	159	89
	29	72	88	84	107	105	82	106	103	163	90
	30	73	90	86	110	108	84	109	105	168	93

CHILDREN'S HOSPITAL NATIONAL MEDICAL CENTER
111 Michigan Avenue, N. W.
Washington, D. C. 20010

ABBREVIATED TEACHER QUESTIONNAIRE

OFFICE USE
Patient No.
Study No.

PATIENT NAME _____

TEACHER'S OBSERVATIONS

Information obtained _____ by _____
                      Month   Day   Year

Observation	Degree of Activity			
	Not at all	Just a little	Pretty much	Very much
1. Restless or overactive				
2. Excitable, impulsive				
3. Disturbs other children				
4. Fails to finish things he starts - short attention span				
5. Constantly fidgeting				
6. Inattentive, easily distracted				
7. Demands must be met immediately - easily frustrated				
8. Cries often and easily				
9. Mood changes quickly and drastically				
10. Temper outbursts, explosive and unpredictable behavior				

COMMENTS _____

FIGURE 2. Abbreviated Teacher Questionnaire

REFERENCES

Aaron, P.G. (1979). A neuropsychological key approach to diagnosis and remediation of learning disabilities. *Journal of Clinical Psychology*, *35*, 326-335.

Abikoff, H., Gittelman-Klein, R., & Klein, D.F. (1977). Validation of a classroom observation code for hyperactive children. *Journal of Consulting and Clinical Psychology*, *45*, 772-783.

Abikoff, H., Gittleman-Klein, R., & Klein, D.F. (1980). Classroom observation code for hyperactive children: A replication of validity. *Journal of Consulting and Clinical Psychology, 48*, 555-565.

Achenbach, T., & Edelbrock, C.S. (1978). The classification of child psychopathology: A review and analysis of empirical efforts. *Psychological Bulletin, 85*, 1275-1301.

American Psychiatric Association, (1980). *Diagnostic and Statistical Manual of Mental Disorders (DSM-III)* (3rd Edition, p. 43). Washington, DC: Author.

Aman, M.G. (1979). Cognitive, social and other correlates of specific reading retardation. *Journal of Abnormal Child Psychology, 7*, 153-168.

Badian, N.E. (1977). Auditory-visual integration, auditory memory, and reading in retarded and adequate readers. *Journal of Learning Disabilities, 10*, 108-114.

Barkley, R.A. (1986). A review of child behavior rating scales and checklists for research in child psychopathology. In R. Prinz (Ed.), *Advances in Behavioral Assessment of Children and Families*, (Vol. 3). Greenwich, CT: JAI Press.

Beitchman, J.H., & Raman, S. (1979). The assessment of childhood psychopathology: The construction of a new self-report psychiatric rating scale for children. *Multivariate Experimental Clinical Research, 4*, 23-31.

Carey, W.B., & McDevitt, S.C. (1980). Minimal brain dysfunction and hyperkineses: A clinical viewpoint. *American Journal of the Disabled Child, 134*, 926-929.

Conners, C.K. (1969). A teacher rating scale for use in drug studies with children. *American Journal of Psychiatry, 126*, 152-156.

Copeland, A.P., & Weissbrod, C.S. (1978). Behavioral correlates of the hyperactivity factor of the Conners Teacher Questionnaire. *Journal of Abnormal Child Psychology, 6*, 339-343.

Copeland, A.P., & Weisbrod, C.S. (1980). Effects of modeling on behavior related to hyperactivity. *Journal of Educational Psychology, 72*, 875-883.

Firestone, P., & Douglas, V. (1975). The effects of reward and punishment on reaction times and autonomic activity in hyperactive and normal children. *Journal of Abnormal Child Psychology, 3*, 201-216.

Reatig, N. (1985). Bibliography on rating and assessment instruments for attention deficit disorder (ADD). *Psychopharmacology Bulletin, 21*, No. 4, 929-1004.

Trites, R.L., Blouin, A.G.A., & Laprade, K. (1982). Factor analysis of the Conners' Teacher Rating Scale based on a large normative sample. *Journal of Consulting & Clinical Psychology, 50*, 615-623.

How Is a Computerized Attention Test Used in the Diagnosis of Attention Deficit Disorder?

Michael Gordon, PhD

ABSTRACT. The computerization of attention tasks has allowed clinicians to incorporate objective data into evaluations for ADD/Hyperactivity. These measures require the subject to cope with demands for self-control and sustained attention. Clinicians should consider practicality, reliability, and robustness of standardization in selecting a computerized test. A well-researched approach to computerized assessment, the Gordon Diagnostic System (GDS), is described. Case histories are presented which illustrate the contribution of GDS data to the evaluation and treatment monitoring of children referred for ADD/Hyperactivity.

The advent of computerized assessment of behaviors associated with Attention Deficit Disorders (ADD) represents another step forward in the effort to define reliable and meaningful diagnostic criteria. These techniques were born of concern about the extensive degree to which diagnostic decisions were founded upon opinion, or upon data from traditional psychological tests of limited relevance to issues surrounding this disorder. Although clinical judgment, behavior rating scales, and clinical interviews are critical to a sophisticated evaluation, each approach has well-documented limitations. In the face of considerable evidence to the contrary, one would be hard put to justify the statement "I know an ADD child when I see one" or the practice of using only rating scales scores or IQ factors to formulate a diagnosis.

COMPUTERIZED ATTENTION TESTS: CONSIDERATIONS

While not intended as magic geiger counters for ADD, computer-based laboratory measures do offer both the researcher and clinician

---

Michael Gordon is Associate Professor of Psychiatry, State University of New York, Upstate Medical Center, 750 East Adams Street, Syracuse, NY 13210.

© 1987 by The Haworth Press, Inc. All rights reserved.

an opportunity to incorporate data derived from a child's actual behavior. Unlike the other clinical methods, laboratory tasks generate objective data about a child's ability to perform in situations tailored to get at the characteristic weaknesses of an ADD child. As such, they generally require a child to sustain attention and control behavior over a period of time and with varying degrees of feedback regarding performance. The tasks are often variants of attentional measures, such as the Continuous Performance Test (Rosvold, Mirsky, Sarason, Bransome & Beck, 1956), in which a child must respond only to a specific combination of symbols in a stream of irrelevant symbols. Many years of research indicate that ADD children fare poorly when required to attend and to develop a strategy for self-control.

Laboratory measures have had a long history in ADD research and a very short one in actual clinical practice. Research versions have traditionally been bulky, expensive, and impractical. There has also existed a certain degree of resistance to using mechanical measures within an evaluation that traditionally has relied heavily on the clinician's judgment. The burgeoning growth of computer technology has solved the technological limitations of earlier electromechanical devices. The increasing popularity of computers in clinical practice has also led to greater comfort with computerized assessment of behavior.

Considerations involved in selecting a computerized assessment technique are essentially identical to those important for the evaluation of any psychological test. The computerization of a measure does not obviate the need for evidence concerning reliability of administration, test-retest reliability, robust standardization, or meaningful studies of validity. Most of the available techniques have enjoyed only limited research investigation of psychometric properties. These procedures have generally been standardized on small groups of subjects, usually children who have been referred to a psychiatric clinic but for reasons other than those associated with ADD. While such samples may suffice for certain research purposes, the clinician is interested more in how a particular child's performance compares to that of peers than to the scores of children referred for a range of psychiatric problems.

Extensive standardization of computerized assessment procedures are particularly critical in the light of the legendary variability inherent in the test performances of children. It is also important to keep in mind that most of the scores generated by these measures are age-related. Without a very substantial normative base, misdiagnosis can result from failure to adjust for a child's developmental status. Another key consideration concerns the practicality of an assessment procedure. If a technique is not designed to fit comfortably into the daily practice of a busy clinician, it will likely fall from use regardless of its diagnostic efficacy. While software-driven assessment programs have been available for many years, they generally have not found their way into clinical practice, often because clinicians have considered them too cumbersome to administer and score. Most practitioners have neither the time nor the computer expertise to struggle with complicated instructions or lengthy procedures. Most of the techniques also fail to meet the clinician's need for portability. Since the majority of those involved in serving children referred for ADD travel among schools and/or offices, they are loathe to carry around bulky equipment or to expend effort in connecting cables and attaching peripherals.

The clinician would, therefore, be wise to select a procedure that can be employed with ease and confidence. Along these lines, the assessment program should be accompanied by clear operating instructions, support materials regarding interpretation, technical data, and active service and research support.

Unfortunately, there are only a handful of available computerized techniques which even begin to approach the criteria mentioned above. Most rely on microcomputers to generate the tasks (Conners, 1980; Klee & Garfinkel, 1983; Greenberg, 1985) in contrast to a self-contained, microprocessor-based unit dedicated to task administration (Gordon, 1982). The advantage of the former strategy is that these software programs tend to be less expensive, more flexible in the varieties of tasks that can be administered, and more amenable to data storage. They also, however, tend to be less transportable, rugged, practical, and reliable in administration. As indicated above, software packages tend to be poorly normed, if normed at all. The one program that has enjoyed some standardization is the Garfinkel Assessment Battery, but this technique has been primarily restricted to research purposes.

GORDON DIAGNOSTIC SYSTEM

One of the most widely-used procedures is the Gordon Diagnostic System (GDS; Gordon, 1982), developed by the author. Although supported by extensive research efforts, the GDS was designed specifically for clinical use. The GDS is a microprocessor-based, portable unit which, without an external microcomputer, allows for the administration of multiple tasks.

The Vigilance Task requires the child to inhibit responding under conditions that make demands for sustained attention. A series of digits flashes one at a time on an electronic display. The child is told to press the button every time a "1" is followed by a "9." The GDS records the number of correct responses, the number of times the child failed to respond to the "1/9" combination (i.e., Errors of Omission), and the number of extraneous button presses (i.e., Errors of Commission). For the testing of younger children, the GDS contains a "1" mode, which requires the subject to press the button every time a "1" appears. The same performance measures are recorded.

The Delay Task requires the child to inhibit responding in order to earn points. Specifically, the child is instructed to press a button, wait a while, and then press the button again. If s/he refrains from responding for at least 6 seconds, a light flashes and a reward counter increments. If the child responds before the interval lapses, then the timer resets and no reward points are recorded. The Delay Task yields three primary scores: the number of responses (button presses); the number of correct responses (i.e., Correct); and the Efficiency Ratio which represents the percentage of correct responses.

The administration of both tasks takes less than twenty minutes. Although normative data were gathered using standard task parameters, the design allows the practitioner to select a wide range of settings. This feature enabled the modification of parameters for the testing of adults as well as very young children. The internal microprocessor generates the tasks and records quantitative features of a child's performance both for the entire session as well as for individual time blocks. In this way, the pattern of a child's performance across the session can be analyzed.

Because the GDS is controlled by a programmable microprocessor, additional tasks can be explored. One such enhancement being studied is a version of the Vigilance Task that assesses the impact of distraction on a child's ability to sustain attention. This Distractibility Task is identical to the standard Vigilance Task except that random digits flash at random intervals on the outer positions of the electronic display. The subject is still required to press the blue button when a "9" comes right after a "1." The only difference is that numbers flash on either side of the center (i.e., relevant) digit. Other enhancements under development are designed to evaluate attention across sensory modalities.

The portability and ease-of-operation of the GDS has allowed for large-scale data collection. The standardization sample is comprised of 1300 boy and girls from 4 to 16 years of age (Gordon & Mettleman, 1985). An additional 1100 hyperactive and nonhyperactive protocols from various subject populations, including deaf, blind, emotionally disturbed, learning disabled, and Spanish-speaking have also been gathered. A series of validational studies has shown that these game-like tasks differentiated accurately between hyperactive and nonhyperactive children from both outpatient and day treatment setting (Gordon, 1979; McClure & Gordon, 1984). In a sample of school-referred children, the GDS distinguished children with ADD from those classified as reading disabled, overanxious, and normal (Gordon & McClure, 1983). Over 20 university and medical center research sites are currently conducting investigations involving the GDS. Most studies concern the effectiveness of the GDS for determining drug responsivity and for evaluating the success of pharmacotherapy (Atkinson, Cohen & Kelly, 1985; Barkley, 1985; Rapport, Dupaul, Kelly & Jones, 1985; Brown, 1986). Others are investigating the use of the GDS for evaluating children with known brain damage, preschoolers considered "at-risk" for impulsive behavior, and the relationship between GDS scores and observational measures of classroom behavior. The GDS is also being used by school systems, mental health professionals, and pediatricians across the country for the clinical evaluation of ADD/Hyperactivity and assessment of therapeutic outcome.

Feedback from clinicians indicates that GDS testing tends to be conducted early in the diagnostic phase in an effort to achieve a more efficient evaluative process. The rationale for this approach is that GDS results can help direct the course of the ensuing evaluation by initially screening for general levels of self-control and attentiveness. For those children who perform within normal limits on the GDS and on brief rating scale measures of hyperactivity, further assessment can be geared toward examining other possible explanations for the child's behavior aside from ADD/Hyperactivity, *per se*. Conversely, the clinician is likely to pursue more intently consideration of a diagnosis of ADD/Hyperactivity for those children whose GDS scores fall in the abnormal ranges.

Practitioners also report that the GDS testing to be most useful in helping to rule out a diagnosis of ADD/Hyperactivity for the many children who are not hyperactive and would best benefit from treatments other than those applied for ADD/Hyperactivity. The procedure is also used extensively in monitoring the effectiveness of pharmacotherapy. Finally, clinicians have consistently reported that, quantitative scores aside, the opportunity to observe a child perform in situations demanding of attention and self-control has been valuable.

CLINICAL EXAMPLES

Presented below are a series of cases in which GDS testing was included within an evaluation (adapted from Gordon, 1984, 1985). While a few case histories, truncated in their description because of space limitations, cannot convey the full impact of objective testing, they can serve to illustrate the sorts of instances where computerized assessment can be meaningful.

Case 1

A nine-year-old boy was referred with complaints of restlessness, noncompliance, fighting with other children, and poor academic performance. While his parents had been recently divorced, these behaviors were longstanding and had been problematic since early childhood. A medical examination by his pediatrician was unremarkable, and he was referred for a psychological evaluation.

Upon initial contact with the psychologist, this youngster appeared inhibited, quiet, and withdrawn, giving no indications of impulsive or hyperactive behavior. However, when confronted with the demands of the Delay Task, the boy was unable to maintain the facade. He achieved an Efficiency Ratio of .44, which falls in the Abnormal Range. His behavior throughout the Delay Task was disjointed, involving much out-of-seat activity and extreme restlessness. While these behaviors in some ways helped him to suppress responding because he was otherwise occupied, he was nonetheless unable to refrain from emitting a large number of unreinforced responses. This pattern of behavior was repeated during the Vigilance Task. He consistently responded to the digit immediately after the appearance of a "1," without waiting to see if it was a "9." His performance was suggestive of an inability to delay, particularly once he had been primed to respond. Following a complete diagnostic evaluation, the youngster was classified as having ADD with Hyperactivity and was placed on a moderate dose of stimulant medication. His academic programming was geared more toward accuracy than speed, and he received resource help to encourage him to modulate his response style. Follow-up contact indicated substantial improvement.

Case 2

Kevin, a 10-year-old boy, was referred for a psychological evaluation by both his school and pediatrician. According to all concerned, Kevin was consistently impulsive, disruptive, inattentive, hyperactive and underachieving. Judging from rating scales and case-history information, his behavior met all of the DSM-III criteria, as well as the more stringent set of markers proposed by Barkley. The age of onset for his problems was before 5-years-11 months, his symptoms were chronic and pervasive, his IQ was well over 70, and ratings of his problem behavior by teachers on standardized checklists were beyond the 95th percentile. His school had demonstrated ample patience with his misbehavior, to the extent that they allowed him to leave the classroom and run the hallways for five minutes when he could no longer contain his energy. At the time of the referral, the school was reaching the end of their capacity to deal with Kevin.

Kevin was administered the Delay Task, Standard Vigilance Task, and the Vigilance Task with Distractors. The Delay Task Efficiency Ratio of .75 was in the Borderline Range for his age. The most noteworthy feature of his performance was the very significant degree of variability across the four Time Blocks. At times, he inhibited perfectly, while at other times his performance approached the Abnormal Range. This marked inconsistency in the context of a profile that shows evidence of the capacity for delay is a pattern found often in children whose problems have a strong emotional component. Unlike the more typical ADD profile in which there is little if any evidence of successful inhibition, the emotionally disturbed child's protocol will tend to show clear swings across the session between adequate and inadequate coping.

Kevin's performance on the standard Vigilance Task also fell in the Borderline Range (albeit the near normal end) for both Correct and Commissions. Again, there were indications of inconsistency of performance in that the middle Time Block, unlike the other two, was normal. As is so often the case, Kevin's behavior during the session was telling. Upon missing the "1/9" combination, he would slap himself in the face and complain that the numbers flashed too quickly for him. The examiner noted that Kevin became so anxious about performing on the task that he had trouble concentrating. His discomfort increased on the Distractibility version of the Vigilance Task in which his performance in all respects was in the Borderline Range.

The GDS evaluation indicated that at times Kevin could be more impulsive and inattentive than age-mates. His poor control, however, did not consistently reach an abnormal range. In fact, he demonstrated a capacity to delay and attend adequately but had difficulty maintaining good performance for reasons that seemed, at least in part, related to emotional issues. His behavior during the testing was that of a boy who became unusually anxious and self-denigrating when he met with frustration or failure.

On the basis of this kind of GDS protocol, a complete psychodiagnostic evaluation is typically suggested. In Kevin's case, a full battery of psychological tests was administered (WISC-R, WRAT, PIAT, Figure Drawings, TAT, Rorschach, Bender, etc.) in addition to clinical interviews with the boy, his family, and his teachers. The overall conclusion from this extensive evaluation was that Kevin's problems of self-control were secondary to a severe and chronic emotional disturbance. The onset of his hyperactivity at age 5 coincided with the gunshot murder of his mother. Because his biological father was unavailable (Kevin was born out-of-wedlock), Kevin was raised by his maternal grandparents. The grandparents were very devoted to the boy but, because of their own fears, suspicions and ambivalence about their custody of Kevin, they seriously limited his contacts with others. They also engaged in what appeared to be bizarre rituals around the memory of the boy's mother. This family history as well as a host of other factors left Kevin with intense fears and a dependent, hostile relationship with his grandparents. As he grew older, he found it increasingly difficult to manage his anxieties and anger. He would become so flooded by stress that he often would have difficulty organizing himself for even simple tasks. Intensive psychotherapy with both Kevin and his grandparents was initiated.

Case 3

A seven-year-old girl was referred to a child development clinic by her teacher for poor school performance, not following instructions, missing assignments, and difficulty with reading. The teacher viewed much of this youngster's behavior as willful and oppositional, and was at her wit's end as to how to help the girl learn. Her parents reiterated the teacher's complaints, and added that they felt the child to be distant and aloof much of the time. During the clinical interview, this youngster was pleasant and cooperative, but gave the impression of either wanting to be somewhere else or of simply daydreaming.

On the Delay Task, she obtained an overall Efficiency Ratio of .82, well within the Normal range. She earned 46 Total rewards. Her responses appeared controlled, orderly, and goal-directed. Obviously, her difficulties did not lie in the area of impulse control. On the Vigilance Task, her difficulty quite clear. Although she produced just one error of Commission, which is well within normal limits, she made 33 errors of Omission, scoring well beyond the mean on this measure.

Although her motivation throughout the task was quite good, she was unable to maintain the degree of preparedness required to perform effectively on this part of the task. We were able, therefore, to identify her main difficulty as a deficit in sustained attention in the absence of impulsivity, which fits the DSM III classification of Attention Deficit Disorder without Hyperactivity.

Following a more extensive evaluation, she was placed on a very small dose of stimulant medication, and her teacher was informed of the results of the testing. Intervention strategies were offered for both home and school, and her adjustment improved in a satisfactory manner.

Case 4

James, a 13-year-old boy of average intelligence, was referred for GDS testing from an adolescent inpatient unit where he was undergoing treatment for a range of psychiatric problems. The precipitant of this second admission to the unit was the boy's attempt to choke himself with a pencil. James had a history of firesetting, suicidal ideation, poor peer relationships, harming animals, and academic underachievement. Full psychological evaluations had pointed to severe pathology across several domains of functioning. He was described as anxious, impulsive, aggressive, depressed, and distractible. In addition to conduct problems, James was considered hyperactive and, upon occasion, had received brief trials of stimulant medication.

Prior to admission, James had resided in a foster home, his fifth placement in as many years. He became unusually unmanageable upon the arrival of a younger foster brother. James had last seen his biological mother three years prior to admission.

While James was considered to be highly impulsive and inattentive, the extent to which those symptoms were secondary to pervasive social and emotional deficits was unclear. Also, there was a need to evaluate treatment with stimulant medication because previous attempts had not been systematically monitored.

James was evaluated with the GDS while off, on, then off medication. (Unfortunately, a double-blind, placebo approach could not be implemented.) James' performance while free of medication fell solidly in the Abnormal Range for both tasks. He demonstrated a clear inability to sustain attention and delay. The practitioner noted that, while the scores were generally typical of the ADD/Hyperactive youngster, the test behaviors were somewhat uncharacteristic. Unlike many ADD/Hyperactive children, James appeared cooperative, anxious, eager to please, and concerned about his performance.

After a week of treatment with methylphenidate, James was retested, and his scores showed marked improvement. His performances on the Delay and Vigilance Tasks were in or near the Normal range, as he demonstrated improved self-control and attentiveness. While he still displayed considerable immaturity, James nonetheless appeared calmer and more focused.

Two weeks later, James was tested once more, this time free of stimulant medication. Even though he boasted that he was an "expert" at these tests, his performance fell back to baseline levels. He was again unable to exert self-control on the Delay Task, and had a high number of commissive errors on the Vigilance Task.

It was suggested that attention deficits were embedded in a constellation of other difficulties, and that attentional aspects of his problems were responsive to pharmacotherapy. The recommendation was made to continue stimulant medication therapy as part of a comprehensive treatment program.

CONCLUSION

The use of computerized assessment in the evaluation of children referred for ADD/Hyperactivity can make for a more accurate, efficient diagnostic process. It allows the practitioner to incorporate data based on the child's actual behavior. Clinicians who pursue this approach need to select a procedure that is practical, well-normed, and energetically supported by research. A legitimate concern is that these computer-generated tasks present a relatively narrow range of demands, stimuli, and response modalities. While the ADD child's impulsivity is viewed as generally pervasive across areas of functioning, no security exists that performance on select computerized tasks will necessarily reflect the overall level of impulsivity for a particular child. In all situations, these objective data must be integrated into a comprehensive evaluation of the sort suggested by Barkley (1981, 1985). As part of a sophisticated clinical protocol, a computerized attention task can significantly enhance the diagnostic process.

REFERENCES

Atkinson, A.W., Cohen, P.C., & Kelly, P.C. (1985, April). *Attention deficit disorder: The effects of ritalin on self-esteem a comparison of ACTeRS teacher scale, Conner's parent scale, and Gordon Diagnostic system in diagnosis and management.* Paper presented at the American Academy of Pediatrics Meeting, Atlanta, Georgia.

Barkley, R.A. (1981). *Hyperactive children: A handbook for diagnosis and treatment.* New York: The Guilford Press.

Barkley, R.A. (1985). *Development of a multi-method clinical protocol for assessing stimulant drug responding in ADD children.* Manuscript submitted for publication.

Brown, R. T. (1986, August). *Controlled of methylphenidate in ADD adolescents.* Paper presented at the 94th Annual Convention of the American Psychological Association, Washington, DC.

Conners, C.K. (1980). *Continuous performance test* [Computer program]. Kensington, MD: Behavioral Medicine, Inc.

Gordon, M. (1979). The assessment of impulsivity and mediating behaviors in hyperactive and nonhyperactive children. *Journal of Abnormal Psychology, 7,* 317-326.

Gordon, M. (1982). *The Gordon Diagnostic System.* DeWitt, NY: Gordon Systems, Inc.

Gordon, M. (1984, 1985). *ADD/Hyperactivity Newsletter.* (Available from Gordon Systems, Inc. P.O. Box 746, DeWitt, NY 13214.)

Gordon, M., & McClure, F. D. (1983). [The assessment of ADD/Hyperactivity in a public school population]. Unpublished raw data.

Gordon, M., & Mettelman, B. B. (1985). Special tables for the *Gordon Diagnostic System.* (Available from Michael Gordon, Ph.D., 750 E. Adams Street, Syracuse, NY 13210).

Greenberg, L.M. (October, 1985). *An objective measure of response to methylphenidate: Clinical validation of the VIRTEST.* Paper presented at the 32nd Annual Meeting of the

American Academy of Child Psychiatry. San Antonio, TX.
Klee, S.H., & Garfinkel, B.D. (1983). The computerized continuous performance task: A new measure of inattention. *Journal of Abnormal Child Psychology, 11*(4), 489-496.
McClure, F.D., & Gordon, M. (1984). The performance of disturbed hyperactive and nonhyperactive children on an objective measure of hyperactivity. *Journal of Abnormal Child Psychology, 12*(4), 561-572.
Rapport, M.D., Dupaul, G.J., Kelly, K., & Jones, J. (1985). *Comparing classroom and clinical measures of ADD-H: Dose response effects.* Manuscript submitted for publication.
Rosvold, H.E., Mirsky, A.F., Sarason, I., Bransome, E.D., & Beck, L.H. (1956). A continuous performance test of brain damage. *Journal of Consulting Psychology, 20*, 343-350.

How Is Playroom Behavior Observation Used in the Diagnosis of Attention Deficit Disorder?

Mary Ann Roberts, PhD

ABSTRACT. The present article describes an observational playroom procedure (Structured Observation of Academic and Play Settings) for evaluating symptomatic behaviors of hyperactive boys in the clinic setting. Previous research has demonstrated that observer-coded measures of on-task behavior and activity level can reliably discriminate hyperactive boys from non-referred boys. Subsequent work, furthermore, has shown that hyperactive boys can be differentiated from aggressive boys and also from boys with a mixed symptom picture of hyperactivity plus aggression. Our studies have generally found that the percentage of time spent on task during the Restricted Academic portion of the observation was the most sensitive variable. Case studies are presented to illustrate the differing pattern of playroom behavior in boys with different types of externalizing disorders. Current research efforts are focused on assessing the robustness of this playroom assessment procedure.

The third edition of the Diagnostic and Statistical Manual of Mental Disorders (American Psychiatric Association, 1980) lists three essential features of Attention Deficit Disorder with Hyperactivity (ADD-H): inattention, impulsivity, and hyperactivity. While behavioral descriptors (e.g., "runs about or climbs on things excessively," and "has difficulty staying seated," APA, 1980, p. 44) are provided for each of these three symptoms, no guidelines are given for measuring the frequency and severity of the symptoms. Thus, each clinician is left to decide how much distractibility is

---

Mary Ann Roberts is a staff psychologist, Division of Developmental Disabilities, Hospital School, University of Iowa, Iowa City, IA 52242. Preparation of this article was supported in part by a National Institute of Mental Health Grant (MH-39037). The author gratefully acknowledges the assistance of Antigone Papavasiliou in referring subjects.

© 1987 by The Haworth Press, Inc. All rights reserved.

pathological and how much and what types of activity constitute overactivity. Most often, it is necessary to rely upon naturalistic observations of the child's behavior reported by parents and teachers. While this information can be systematically elicited on standardized questionnaires, the data, and ultimately then the diagnosis, depend largely on the normative standards of the individuals completing the questionnaires. The task is further complicated because attention span and activity level vary with age, and the extent to which parents and teachers modify their judgements based on the developmental level of the child is unclear.

For the diagnosing clinician, observing the reported behavior problems first hand is particularly convincing and perhaps has the greatest face validity of all assessment methods. Unfortunately, as suggested in DSM-III (APA, 1980), a child who truly exhibits ADD-H may behave in an unremarkable fashion in the clinician's office (Sleator & Ullmann, 1981). Behavioral observation of the child in the naturalistic environment, particularly in school, presents an appealing alternative (Abikoff, Gittelman & Klein, 1980). Unfortunately, school settings vary widely. Adequate normative data are difficult to obtain because significant aspects of the situation (e.g., size of the room, daily schedule, individuals present) cannot be standardized. Furthermore, classroom observations are generally impractical for the clinician in terms of time and expense.

In order to provide an objective source of behavioral data for diagnosis of ADD-H children, I (Roberts, 1979; Roberts, Ray & Roberts, 1984) and several of my colleagues (Milich, Loney & Landau, 1982) have developed a structured playroom observation procedure for the clinic setting. Our playroom procedure has its origins in a method proposed by Routh (1980). In addition to the Free Play session used by Routh, we have added an observation period designed to approximate a school or task-oriented situation (Restricted Academic). This combination of Free Play and Restricted Academic periods is called the Structured Observation of Academic and Play Settings (SOAPS), and a detailed manual summarizing the procedures involved is available from the author (Roberts, Milich & Loney, 1985). What follows in the rest of the chapter is a description of this playroom procedure and how an observation is conducted. Next, the results of our studies using this procedure are summarized. Finally, several illustrative case examples are presented.

PLAYROOM PROCEDURES

The SOAPS playroom procedure consists of two 15-minute observation periods, the Free Play and Restricted Academic settings. In the Free Play period, the child is allowed to play freely with any of the toys available in the room. During the Restricted Academic period, the child is instructed to complete a set of worksheets and to refrain from playing with the available toys. The behaviors coded by observers are those traditionally associated with Attention Deficit Disorder with Hyperactivity, specifically including measures of activity level and attention to task.

In the course of conducting our research studies on this observation procedure, it has been necessary to equip several playrooms at different sites. As the physical setting in each case was slightly different, we have established general guidelines for equipping an observational playroom for clinical use (Roberts, Milich & Loney, 1985). The room must be large enough to accommodate four child-sized tables and chairs, with each table and chair spaced equidistant from the center of the room. The floor is divided into grids, and the number of grids crossed during an observation period is used to provide one measure of motor activity. During the Free Play period each table contains the same set of five toys: a water toy, a friction car, drawing materials, an Etch-a-Sketch, and a set of Tinker Toys. For the Restricted Academic setting, all of the Free Play toys are removed, and a set of five worksheets with pencils are placed on three of the four tables. The fourth table contains a set of interesting distractor toys which the child has not seen during the Free Play period. Typically, we have included some type of space station toy and a pocket-sized video game. All other materials in the playroom should be removed if possible (e.g., all other play materials). It should also be equipped in such a way that the child could remain alone in the room safely for 15 minutes.

In our research studies, each child wore a wrist and ankle actometer. Using actometers to measure locomotor activity has both advantages and disadvantages. While one incurs the initial expense of purchasing the actometers, it should also be noted that actometer scores can be obtained without an observer having to be present at each playroom session. Considerations in constructing and using actometers are detailed in the SOAPS manual.

At the beginning of the SOAPS observation, the child is brought to the playroom and given instructions for the Free Play or Restricted Academic period. For the Free Play period, the child is told by the examiner that he/she may play with any of the toys available for the next 15 minutes. The examiner then leaves the room and returns when the 15-minute observation period is completed. For the Restricted Academic period the child is told that the room is now set up like a classroom with three desks and an activity table. The examiner reviews the worksheet instructions, directs the child not to play with the activity table toys, and once again leaves the room for 15 minutes. Approximately 45 minutes is necessary to complete the entire evaluation with both observation periods. Children undergoing the SOAPS evaluation typically enjoy the playroom activities and often consider it to be a preferred part of their time at the clinic.

Observation Procedures

While the child is in the room during the Play and Academic periods, an observer records the child's behavior from the other side of a one-way observation window according to previously defined codes: number of grids crossed, time spent out of seat, time spent on task, number of attention shifts, time spent in restless, fidgety movement, and time spent vocalizing. Given that some clinicians may not have access to a playroom with an observation window, we are currently studying the effect of an adult's presence in the playroom during the observation. Unfortunately, preliminary results suggest that an adult's presence in the playroom may significantly influence the child's behavior.

Studies on SOAPS and Case Examples

To date, several diagnostic studies have been conducted using the SOAPS procedure with boys 6 to 12 years of age. In two separate studies, data from the Restricted Academic period successfully discriminated hyperactive and non-referred boys with 88% and 90% accuracy (Roberts, Ray & Roberts, 1984). The ADD-H children were less attentive, less productive (worksheet scores), and exhibited more locomotor activity. These results demonstrated that when the SOAPS procedure was used, boys diagnosed as ADD-H did indeed behave differently from non-referred peers in the clinic playroom. However, rather than discriminating normal from pathological behavior, the clinician typically confronts the problem of differentiating among various behavior disorders of childhood. With regard to ADD-H, it is particularly difficult to discriminate boys with this disorder from boys with other externalizing disorders, such as aggressive conduct disorder. Nevertheless, with regard to long-term prognosis, Loney's research (Loney, Kramer & Milich, 1981) underscores the importance of accurately distinguishing among these groups at the time of referral.

In another study, Roberts (1979) demonstrated that the SOAPS procedure could be used to differentiate with a high degree of accuracy groups of boys with purely hyperactive (ADD-H) symptomatology, boys with purely aggressive symptomatology, and boys with a combined disorder (hereafter called hyperactive plus aggressive). On the basis of extreme scores on parent and teacher rating scales and psychiatric chart ratings, boys referred to a child psychiatry clinic were assigned to one of three groups: hyperactive only, aggressive only, or hyperactive plus aggressive. A discriminant function based on SOAPS data from the Restricted Academic period correctly classified 86% of the boys in these three groups. In a subsequent study, Milich et al. (1982) provided conceptual replication of these findings. Over the course of these studies, the Restricted Academic period has proven to be the most powerful situation for eliciting the behavioral symptomatology of ADD-H boys. Furthermore, of all the variables coded in the SOAPS procedure, the attentional variable of time on-task during the Restricted Academic period has produced the greatest degree of discrimination among purely hyperactive, purely aggressive, and hyperactive plus aggressive boys. Figure 1, taken from Roberts (1979), displays the distribution of individual subjects for this measure. For purposes of comparison, the distribution of scores from a sample of nonreferred boys is also provided in Figure 1. Inspection of Figure 1 reveals that there is relatively little overlap among clinic groups on this attentional measure.

The purely aggressive boys were quite capable of sustained attention to task and generally did so with little difficulty (80% to 100% on-task). In the playroom, their behavior could be described as quite goal-directed. These boys were aware that they were undergoing a psychiatric evaluation, and they were careful to exhibit exemplary behavior. Allen, a 10-year-old fourth-grader enrolled in a regular classroom, typifies this pattern. Based upon behavior rating

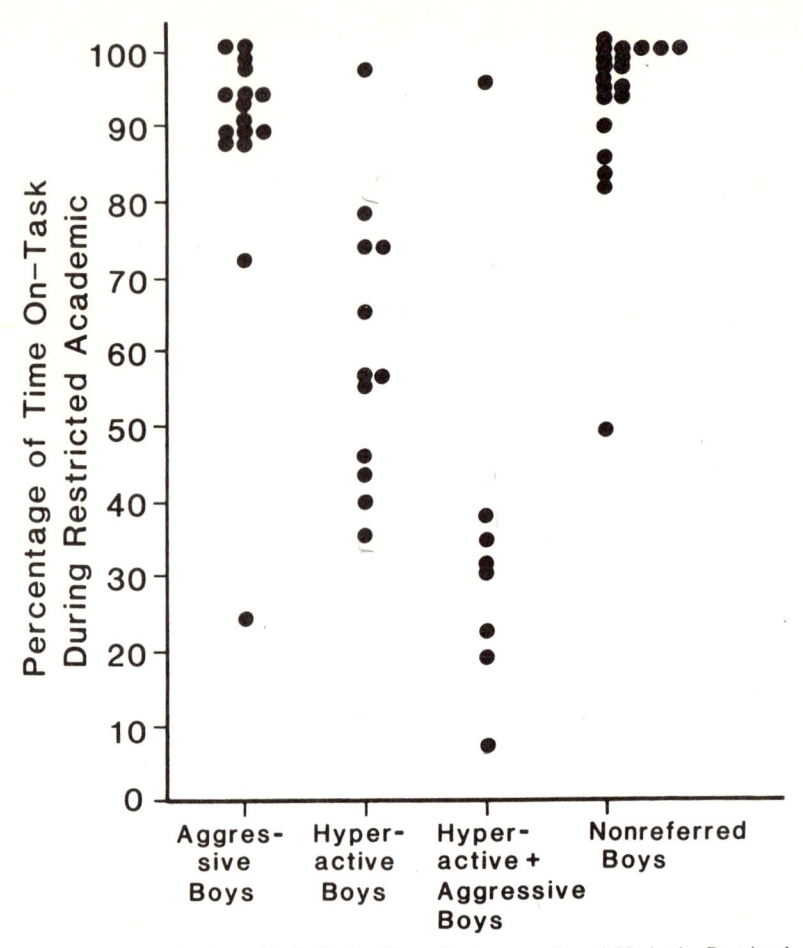

FIGURE 1. Distribution of individual subjects for the on-task variable in the Restricted Academic period.

scales and a parent interview, Allen's mother reported noncompliance and defiant behavior to be of primary concern. Allen was described as being resentful toward his parents and siblings, and he took particular delight in destroying their prized possessions. Although he had not been caught, he was also believed to have taken some small items from a local store. In school, Allen's teachers were not concerned about his academic capability. He generally performed at an average level, although his teachers felt he could do better if he tried harder. His behavior in class was relatively unremarkable, although Allen's fourth grade teacher reported that he looked sullen at times. On the playground, Allen had been observed to get into a number of fights. He teased his peers and verbally intimidated children both younger and older than himself. In contrast to the teacher's description of Allen's sullenness in the classroom, he appeared to be cheerful and outgoing on the day of the playroom evaluation. He was cooperative during psychological testing to the point where the testing technician was puzzled as to why Allen was being seen in a psychiatric clinic. During the Restricted Academic portion of the SOAPS assessment, Allen attended to the worksheets 93% of the observation period. This level of performance is well within the range of scores produced by aggressive boys (See Figure 1). Allen glanced at the distractor toys on the activity table on two occasions, but never left his seat to play with them. Throughout the evaluation day, Allen appeared to be able to control his behavior and generally created a favorable impression among those individuals who worked with him.

In contrast to the behavior of aggressive boys, such as Allen, purely hyperactive boys tended to exhibit noticeable difficulty in attending to the worksheets during the Restricted Academic period (40% to 80% on-task, See Figure 1). During the observation, the behavior of these boys is best described as well-intentioned. Hyperactive boys tried diligently to exhibit the expected behavior. Clinical observation of the individuals in this group indicates that, despite their best efforts, the hyperactive boys were rather inefficient in their attention to task. In this regard, the behavior of Henry, an 8-year-old third-grader, is fairly typical. Based upon rating scales and a structured interview, Henry's mother reported excessive daydreaming, poor academic performance, immaturity, and poor peer relationships to be of primary concern. Henry's mother indicated that he seemed unable to concentrate on anything, even his favorite play activities, for more than a minute or two. He was distractible to the point where he appeared not to hear instructions given to him. Henry's mother was not particularly concerned about oppositional or defiant behavior. At school, Henry was reported to be working below the third grade level, especially in the areas of math and handwriting. He was characterized by his teacher as disorganized and inattentive. He failed to complete assignments and would frequently guess at the answers just to have the work finished, rather than considering his responses thoughtfully. On the playground,

Henry preferred to play with the first graders rather than his classmates, who often excluded him from their activities due to his silliness and immature behavior. During the clinic evaluation day, Henry was cooperative and compliant with the examiner. In a structured interview, Henry complained that he didn't have many friends, but was unable to explain why this was so. Henry appeared to be eager to please the examiner, but often tried to jump ahead to the next task before completing the one at hand. During the Restricted Academic period of the SOAPS assessment, Henry attended to the assigned worksheets only 60% of the time. Following the examiner's departure from the playroom, Henry initially sat down and completed a few items on the first worksheet. Shortly thereafter, he was drawn to the activity table to take a quick look at the toys. After only a few moments playing with the toys, Henry ran back to complete a few more worksheet items. Rather quickly, however, he was once again inextricably drawn back to the activity table, and the cycle started over. Despite Henry's good intentions and his eagerness to please, he worked in a disorganized fashion and was unable to consistently control the focus of his attention.

While the hyperactive boys, such as Henry, exhibit intermittent control over their behavior, the hyperactive plus aggressive boys typically manifest even less capacity to regulate their behavior and appear unconcerned over their lack of responsiveness to external authority. Boys in this group exhibited the greatest difficulty attending to the worksheets during the Restricted Academic period (0% to 40% on-task, See Figure 1). In the playroom, their behavior could be described as rather uninhibited. Matt, an 8-year-old third-grader, exemplifies this behavior pattern. During a parent interview, Matt's mother complained in a defeated and distressed manner that her son displayed a variety of severe behavior problems, including frequently hitting and teasing his two-year-old sister, throwing frequent temper tantrums, and threatening to run away. When punished for his misbehavior, Matt failed to express guilt or remorse and, in fact, dared his parents to punish him again. His behavior at school was so difficult to manage that he was being considered for placement in a special class for behavior problem children. He was openly defiant to adults at school and bullied his peers on the playground. When confronted with his misbehavior, Matt would blame other children or make up improbable alibis. When not on stimulant medication, Matt did virtually no work in school. Prior to the clinic evaluation, he had been withdrawn from medication.

On the day of the testing, he had to be closely supervised lest he run off. During the psychological testing he was irritable and oppositional. At the start of the Restricted Academic period, Matt informed the examiner that he had no intention of completing any worksheets. Despite the examiner's presence, Matt ignored the instruction to sit and do the worksheets and walked directly over to the activity table. Only after the examiner left the room did he return to the worksheets for a brief period. In total, Matt was on-task only 35% of the time during the Restricted Academic period. He tried to leave the playroom on several occasions during the observation, and attempted to take part of the distractor toy home in his pocket. Matt demonstrated very limited capacity to control his behavior, and unlike Henry, the hyperactive boy, Matt appeared not to care.

In our current research we are investigating the potential of the SOAPS procedure for monitoring response to treatment with stimulant medications in hyperactive and hyperactive plus aggressive boys. For this purpose, Matt was evaluated with playroom observations on two subsequent occasions. Ninety minutes prior to the first follow-up observation, Matt was administered Ritalin in the dosage of .3 mg/kg of body weight. On this dose of medication, Matt continued to be oppositional, extremely impulsive, and required close supervision by clinic personnel. In the Restricted Academic portion of the playroom observation, he was on-task only 30% of the time. During a week-long trial at this dosage, Matt's parents reported that his behavior remained extremely disruptive. Subsequently, Matt was reevaluated after receiving .7 mg/kg of Ritalin. On this larger dose, Matt was pleasant to the examiner, did not attempt to run off as he had on previous occasions, and complied with instructions. In the Restricted Academic portion of the SOAPS observation, Matt attended to the assigned worksheets for 75% of the time. He worked carefully on the assigned sheets and was quite productive. The behavior ratings completed by Matt's parents during the week-long trial on the higher dose indicated marked improvement in a variety of problem behaviors. Nevertheless, despite this good outcome which prompted Matt's physician to prescribe .7 mg/kg on an ongoing basis, it should be noted that the proportion of time Matt spent attending to the worksheets in the playroom did not reach the level shown by nonreferred boys. Similar findings from additional case studies presently suggest that the SOAPS assessment may prove useful in evaluating or predicting response to treatment.

SUMMARY

In summary, the research described above indicates that the SOAPS procedure provides an objective means of assessing the behavioral symptoms of boys with Attention Deficit Disorder with Hyperactivity in the clinic setting. The procedure is relatively brief, can be conducted by trained para-professionals, and should prove feasible in most clinical settings. In addition, this method has considerable potential for distinguishing behaviorally homogeneous subgroups of children with externalizing disorders. However, it must be acknowledged that the SOAPS procedure provides only one source of data which must be integrated with other types of information prior to making a clinical diagnosis. Our current research efforts are directed towards evaluating the robustness of the SOAPS method with different playrooms across multiple sites.

REFERENCES

Abikoff, H., Gittelman, R., & Klein, D. F. (1980). Classroom observation code for hyperactive children: A replication of validity. *Journal of Consulting and Clinical Psychology, 48*, 555-565.

American Psychiatric Association. (1980). *Diagnostic and statistical manual of mental disorders* (3rd ed.). Washington, DC: Author.

Loney, J., Kramer, J., & Milich, R. (1981). The hyperkinetic child grows up: Predictors of symptoms, delinquency, and achievement at follow-up. In K. Gadow & J. Loney (Eds.), *Psychosocial aspects of drug treatment for hyperactivity*. Boulder, CO: Westview Press.

Milich, R., Loney, J., & Landau, S. (1982). Independent dimensions of hyperactivity and aggression: A validation with playroom observation data. *Journal of Abnormal Psychology, 91*, 183-198.

Roberts, M. A. (1979). A behavioral method for differentiating hyperactive, aggressive, and hyperactive plus aggressive children (Doctoral dissertation, University of Wisconsin-Madison). *Dissertation Abstracts International, 40*, 3962.

Roberts, M. A., Milich, R., & Loney, J. (1985). *Structured observation of academic and play settings (SOAPS): Manual*. (Available from Mary Ann Roberts, Ph.D., Hospital School, University of Iowa, Iowa City, IA 52242).

Roberts, M. A., Ray, R. S., & Roberts, R. J. (1984). A playroom observational procedure for assessing hyperactive boys. *Journal of Pediatric Psychology, 9*, 177-191.

Routh, D. K. (1980). Developmental and social aspects of hyperactivity. In C. K. Whalen & B. Henker (Eds.), *Hyperactive children: The social ecology of identification and treatment* (pp. 55-74). New York: Academic Press.

Sleator, E. K., & Ullman, R. K. (1981). Can the physician diagnose hyperactivity in the office? *Pediatrics, 67*, 13-17.

PROGNOSIS

What Happens to "Hyperactive" Preschoolers?

Emily K. Szumowski, MS
Linda J. Ewing, RN, BS
Susan B. Campbell, PhD

ABSTRACT. Overactivity, limited attention span, and impulsivity, the diagnostic indicators of Attention Deficit Disorder, are often characteristic of preschool children. Longitudinal research indicates that, as a group, preschoolers showing high levels of these and related problem behaviors are likely to continue to have more difficulties at school-age. The prediction of outcome for individual children, however, is not straightforward. It is important to examine each child's behavior in a variety of settings, to consider parental perceptions and behaviors, and to view the family in its social context. We present case examples from our research illustrating differential outcomes for "hyperactive" preschool children.

CLINICAL ISSUES IN THE DIAGNOSIS OF PRESCHOOLERS

With children of any age, it is often difficult to distinguish between normal and deviant behavior because behaviors change dramatically with development. With preschool children, however, the

---

Emily K. Szumowski and Linda J. Ewing are Graduate Research Assistants, Department of Psychology, University of Pittsburgh. Susan B. Campbell is Professor of Psychology, Department of Psychology, University of Pittsburgh, 4015 O'Hara Street, Pittsburgh, PA 15260. The research project described in this chapter was supported by NIMH Grant No. R01 MH32735 to Dr. Campbell.

© 1987 by The Haworth Press, Inc. All rights reserved.

difficulty is increased because developmental change is particularly rapid and because typical behaviors are often annoying to adults, making them even more difficult to distinguish from truly deviant behaviors (Campbell, 1985). Some preschool children show extreme forms of these typical, but annoying, behaviors, which may be replaced by more mature, appropriate behaviors as development proceeds. When these behaviors include high levels of motor activity, impulsivity, and distractibility, parents often wonder whether children are merely going through a difficult developmental phase or whether they are showing early signs of hyperactivity. While the answer should gradually become clearer as the child grows older, an accurate prediction of likely outcome during the preschool years would help considerably, either by allaying parents' fears or by steering them toward appropriate intervention programs.

The few existing follow-up studies of preschool children have generally suggested that, as a group, children who showed higher levels of problem behavior during the preschool years were more likely to continue to show higher levels of problem behavior two to seven years later (e.g., Campbell, Ewing, Breaux & Szumowski, 1986; Fischer, Rolf, Hasazi & Cummings, 1984; Richman, Stevenson & Graham, 1982). For example, Fischer et al. (1984) obtained parent ratings of the behaviors of 541 children at 2 to 6 years and again at 9 to 15 years. The children who were rated as showing severe externalizing behaviors during the preschool years were significantly more likely to be rated in the clinically disturbed range at follow-up. The authors cautioned, however, that although their results indicated significant stability within the *group* of children showing externalizing problems, the degree of stability was relatively modest; discontinuity rather than continuity was more typical for individual children. What is needed is an understanding of the factors that influence stability vs. change in young children's problem behaviors.

Although we do not yet have a set of foolproof criteria that can reliably distinguish extremes of typical behavior from precursors of hyperactivity in preschool children, it is clear that many factors must be considered. Besides the age of the child, objective qualities of the problem behavior must be weighed, including its duration, frequency, severity, pervasiveness across situations, and association with a constellation of other problematic behaviors. Parental behaviors and the environmental context must also be considered. In addition, because these objective data must be collected from adult observers (typically, parents and teachers), subjective factors influencing the informants' reports must be considered, including their knowledge of child development, expectations, mood, and tolerance for rambunctious, but age-appropriate, child behavior. In our ongoing study of the behavior of preschool children (Campbell, Breaux, Ewing & Szumowski, 1985; Campbell, Szumowski, Ewing, Gluck & Breaux, 1982), several of these factors significantly predicted the persistence of problems. Fifty percent of the children rated by their mothers as showing behavior problems at age 3 continued to show hyperactive and/or aggressive behavior at age 6, despite differential attrition from among the most disorganized families in the problem group. The significant predictors of persistent behavior problems included child characteristics (higher levels of hyperactive and aggressive behavior as rated by mothers on questionnaires and higher levels of activity, inattention, and negative behavior observed during laboratory sessions), parent characteristics (higher levels of negative maternal behavior observed in the laboratory during play), and social context variables (lower socioeconomic status and higher levels of family stress). Although this combination of factors was uniformly associated with a poor outcome (i.e., children high on all of them did poorly), in most cases the influence of child characteristics, parent characteristics, and social context factors was more variable. That is, these various factors appeared to be more important in some cases than in others.

In the next few pages, we will describe this longitudinal study briefly and then provide two case descriptions to illustrate how variations in several of the above parameters appeared to lead to different outcomes. Some children qualified for a DSM-III diagnosis of ADD with Hyperactivity at school entry, others displayed a variety of symptoms, but were not clearly ADD, and still others appeared to overcome their early difficulties.

A LONGITUDINAL STUDY OF BEHAVIOR PROBLEM PRESCHOOLERS

Our study was designed to investigate the development of attention, self-control, and activity level in young children. We obtained a sample of parent-referred children (N = 46) who were perceived to be difficult to manage and a comparison group of children

(N = 22) whose parents felt them to be free of problems. Children were recruited from pediatricians' offices, preschools, and "mothers-day-out" groups.

Participating families were visited in their homes, where mothers were interviewed to assess their children's current behavior, birth history and early temperament, parental disciplinary styles, and family relationships. Parents also completed several questionnaires assessing their perceptions of their children's activity, attention, aggression, and anxiety. During the home visits, the children were administered the Stanford-Binet Scale of Intelligence. Each family also made two visits to our clinic playroom, where the child was observed during free play, during structured play with mother, and during the administration of structured tasks. These observations assessed children's attention, gross motor activity, and impulsivity, as well as the quality of mother-child interaction. Finally, observations were conducted on the 33 study children who were enrolled in preschool.

Sixty-four families completed the assessment. Parents were then invited to return for a comprehensive feedback session. Parents from the problem group also were invited to participate in an eight-week parent training group. Children and their mothers were seen for follow-up assessments at ages 4 and 6. Interview, questionnaire, and observational data were again obtained. At age 6, school observations were also conducted and teachers completed questionnaires. As each study child reaches age 9, additional interview and questionnaire data are being collected.

THE "HYPERACTIVE" CHILD AS A PRESCHOOLER

Mrs. M. was referred to the research project by a local psychologist, who had seen her son David, age 3 1/2, intermittently since he was 2. The psychologist had devised a behavior management program for David and his parents that focused on his overactive and non-complaint behavior. Despite this, a recent escalation in his unprovoked aggressive outbursts had resulted in his expulsion from preschool. It was this event that precipitated Mrs. M.'s request for further help. According to Mrs. M., David's short attention span, high activity level, extreme impulsivity, violent temper, and aggressive episodes were seemingly no longer affected by the behavioral strategies that had been somewhat effective for at least a year. She was, therefore, quite desperate for additional help.

Almost all the parents who telephoned our project with concerns about their preschoolers described a constellation of behaviors that included a short attention span, high activity level, and poor self-control. A number were also described as aggressive and/or noncompliant, underlining the difficulty of separating possible early signs of hyperactivity from those of oppositional or conduct disorders. It was clear, however, that the actual behaviors exhibited by individual children ranged from developmentally age-appropriate, but annoying, behavior to very difficult, potentially clinically significant behavior. Further, in a significant number of children, the target behaviors were present to a greater or lesser extent contingent on the setting (laboratory, preschool, or home) and the demands of the situation (e.g., the degree of structure). Thus, although these children sounded similar according to parental reports of manifest behavior, they varied considerably in the severity of their problems and in how consistently they demonstrated them across settings and situations.

During the home interview with Mrs. M., it was learned that David and his younger brother lived in an intact lower middle class family. His high school educated parents both worked at semi-skilled jobs, but had arranged their schedules so that one parent was always home with the children. Other caretakers had been unable to cope with David's difficult behavior. Mrs. M. described her marriage as happy, but reported that David had been a persistent source of stress and conflict for the family virtually since his birth.

Family stress, including separated or divorced parents, financial or job instability, frequent arguments, and frequent absences of one parent have been associated with less desirable outcomes at school age among children showing externalizing behavior problems as preschoolers. Thus, in order to identify environmental factors that may be contributing either to difficult child behavior or parental perceptions of problem behavior, it was important for us to assess marital satisfaction, family stability, and family support systems. In this case, David himself seemed to be the major source of family stress in an otherwise stable home.

It is also important to obtain a thorough history of the young child, including information about the mother's pregnancy, the birth itself, and the first two years of life. This information can provide valuable insights into the accomplishment of normal developmental milestones and the age of onset of difficult behaviors. It may also suggest that parents have faulty expectations about normal development or exaggerated or distorted perceptions of problems. Although the usual cautions apply when obtaining information retrospectively, the value of this early developmental history cannot be underestimated.

In describing David as an infant, Mrs. M. related that although he was the 6 lb., 3 oz. product of a 42-week pregnancy, he experienced respiratory difficulties immediately after birth and required low doses of oxygen in an isolette for several days. While he seemingly recovered rapidly from his breathing difficulties, he was reported to be a colicky, irritable infant who resisted cuddling and was difficult to soothe. He walked at 8 1/2 months of age, and maintained a very high activity level from that time on. As he entered toddlerhood, his lifelong sleep difficulties persisted, and head banging and rocking became part of his bedtime ritual. In addition, David's parents began to note his seemingly purposeful destruction of toys, excessively reckless climbing, biting, and inability to attend to any activity for more than a few fleeting moments. The birth of his younger sibling at about this time only served to exacerbate his problems further. Shortly after this, David's parents first sought help. They were subsequently taught to implement behavior management strategies which they did with some positive impact on David's behaviors until the present referral.

In David's history, there are indications of difficulties almost immediately after birth. Although his perinatal respiratory problems remitted quite rapidly with only minimal intervention, he is described as a temperamentally difficult infant who resisted cuddling and soothing. Feeding and sleeping problems persisted throughout the first year, and as he added walking to his repertoire at 8 1/2 months, it became another source of distress for his parents who were unable to contain his activity. Although only a small percentage of children exhibiting similar problems during the first year of life continue to have difficulties when followed prospectively, the situation is quite different when data are obtained retrospectively. It is not unusual for mothers of toddlers or preschoolers showing a cluster of externalizing behaviors (overactivity, impulsivity, inattention, aggression, non-compliance) to report a very difficult infancy. While these maternal reports may be influenced somewhat by current difficulties controlling their children's behavior, it also seems clear that they have some validity. This seemed to be the picture emerging with David.

At the home visit, our first contact with David, he was noted to have a high level of activity, described as frenetic, and an extremely limited attention span. He virtually ricocheted from one toy to another while awaiting the administration of the Stanford-Binet. Although it was difficult to engage him in structured tasks because of his unwillingness (inability?) to maintain eye contact, his limited attention span, and persistent irritability, the testing was completed. David scored in the low-average range, but in view of his distractibility and limited cooperation, there was some doubt about the validity of the score.

During both visits to our laboratory playroom, David's short attention span and high activity level were again apparent, as were a moderate amount of irritability and non-compliance. For example, during free play, David flitted rapidly from one toy to another, but failed to become engaged in a constructive or sustained play activity. As with the Binet, however, David could be coaxed to complete structured tasks if he was provided with frequent redirection and praise. David's mother was observed to provide him with both clear limits and liberal praise throughout their interactions. She was frequently and appropriately firm, but also warm and supportive.

David was not observed at preschool since he had been expelled prior to his enrollment in our study. But that incident and teacher accounts of his aggression toward his peers provided additional data which further validated the pervasiveness of his problems.

Observing the child in different settings (home and preschool), and gathering information about current behavior from both parents *and* teachers help to provide a comprehensive picture of actual behavior. Such an approach also provides important information on which situations and/or individuals elicit the difficult behaviors. It is important to determine if the problem behaviors are situation-specific (e.g., only at home with mother or only at preschool), or whether the same behaviors are a problem for parents, grandparents, babysitters, and teachers alike. Children whose difficulties are manifest more pervasively are more likely to have a *bona fide* problem, one that is unlikely to be only "in the eyes of the beholder." These are the children whose problems are less likely to be transient; rather, they are more likely to show persistent problems at school entry and beyond.

At this point, it was clear to us that David's behavior problems were pervasive and were creating quite a bit of distress for his family. It was felt that David's parents could benefit from the eight-week parent support/education group that was offered in conjunction with our research. They willingly participated. They were helped to implement several different behavior management strategies and to target the most troublesome behaviors for intervention. Even though David's parents were by now quite sophisticated in the use of the principles of behavior management, they benefitted not only from the close monitoring we could provide as they worked on specific, high frequency, resistant behaviors, but also from the support and encouragement of the other parents and group leaders. Parents of truly difficult youngsters often experience a concomitant problem of being isolated and rejected by *their* peers. Many parents report feeling judged as incompetent by other parents and teachers despite the extraordinary demands placed on them by their behaviorally handicapped child. The support these parents can provide for one another in the context of a learning environment is an additional benefit which we had initially underestimated.

At our first follow-up one year later, David (age 4 1/2) was managing relatively well in a specialized preschool program. In this highly structured setting, peer aggression seemed to be less of a problem, although Mrs. M. reported that David continued to be aggressive toward his younger brother. David's tantrums had decreased in frequency, but had increased in intensity and occurred when he was confronted with even low levels of frustration. He also remained highly active, inattentive, and non-compliant.

When David was 5, his pediatrician suggested a trial of methylphenidate and he appeared to respond favorably, but not dramatically, to medication.

By age 6, David was in the public kindergarten, but there was some question whether he would remain. His problem behaviors including short attention span, impulsivity, and hyperactivity were severe enough that, according to both teacher and parent reports, David met DSM-III criteria for Attention Deficit Disorder with Hyperactivity. He had been referred by his kindergarten teacher to the school psychologist for evaluation of both his behavior problems and his suspected learning disabilities.

David is clearly representative of children whose early symptoms of hyperactivity, impulsivity, short attention span, and aggressivity were clinically significant, severe, and pervasive from early on. Further, despite early intervention, problems persisted from preschool to school age and, by age 6, there was no doubt that David met DSM-III criteria for ADD with Hyperactivity (as well as for Oppositional Disorder). Further, David's very severe behavior problems have persisted in spite of his parents' ability to provide consistency, structure, and well-established limits, an intact, loving family environment, and low levels of additional family stress.

It should be noted, however, that the manifestations of David's problems have not remained static. Rather, they seem to have appeared first as difficulties with mood, consolability, and self-regulation in infancy; as overactivity and recklessness in early toddlerhood; as a short attention span, difficulty playing alone, tantrums and aggressive outbursts in the preschool years; and, more recently, as age-inappropriate but excessive defiance and tantrum behavior in addition to continued overactivity, poor impulse control, limited attentional capacities, and suspected learning disabilities.

Parenting effects and social context effects appear to be less salient and to be more unrelated to the persistence of David's difficulties than they are with other children we have seen. Rather, it seems likely that David has a biological vulnerability which is contributing to the continuation of problems and placing him at continued high risk for future problems.

Annie provides a contrasting example. Mrs. J. called the project when her daughter Annie was two years, five months old after seeing our poster in her pediatrician's office. During the initial telephone contact, she expressed concerns about Annie's high energy level, tantrums, sleep problems, and fearfulness. At the home interview, she also complained about Annie's low frustration tolerance, impatience, fussiness, and difficulty playing alone or sustaining attention, although she reported that Annie could amuse herself for up to 20 minutes and enjoyed watching Sesame Street, both of which suggested that her attention span was well within the typical range for a 29-month-old.

Annie was the first-born child of college-educated parents. Her father held a managerial position. Annie's mother had stopped working just prior to her birth and was home with her full time. The marriage was reported to be stable with occasional disagreements over Annie. Annie was the product of a full-term pregnancy which was complicated by persistent severe nausea throughout. Mrs. J. reported that she had had concerns about Annie since early infancy. She recalled that Annie had slept less than she had expected her to (although she also indicated that she had slept for six or seven hours through the night from the age of six weeks!), had experienced feeding problems, was difficult to soothe, and did not like to be cuddled. Further questioning, however, did not clearly substantiate these patterns.

Although at this point in the history, there seemed to be many similarities between David and Annie, it became increasingly clear that Mrs. J. was a highly anxious woman with doubts about her own maternal competence. She appeared to want to keep Annie as an infant, dressing her in "baby" clothes, maintaining bottle feeding, sitting her in a high chair, and giving her little exposure to peers. She acknowledged her own uncertainty about how to manage Annie's behavior, and reported that she had tried reasoning, "threatening," yelling, smacking, and time out, but had been frustrated by her inability to enforce limits. This had led to frequent struggles between Annie and her mother, many of which Annie "won" with her persistent tantrums. Whereas Mrs. J. perceived Annie's difficult behavior as purposely provocative, Mr. J. found her much easier to handle and saw her problems as less severe.

Even at home, Annie was quite distressed at separating from her mother during the administration of the Stanford-Binet. She sought frequent contact with her mother and was generally irritable and whiny, refusing several tasks. In the playroom, Annie was active, impulsive, and demanding, as well as frequently off-task and out of her seat during structured tasks. Annie's mother was simultaneously observed to be intrusive and naggy, lacking in warmth, and ineffective in setting limits or enlisting Annie's cooperation, even during play. Overall, the interaction was fraught with tension and conflict over who was in control, and neither mother nor daughter exhibited positive affect.

Although Mrs. J. described Annie as a difficult infant, and we did observe the overactivity and non-compliance she reported, it was hard to assess what actually was going on here. The inconsistencies in Mrs. J.'s reports of Annie's behavior, and her inability to set firm limits or to provide opportunities to facilitate Annie's development were all indications of problems with mothering. Annie's behavior, while it might have reflected early signs of behavior problems, was also quite consistent with the strivings for autonomy and independence characteristic of the two-and-a-half-year-old.

When Annie was seen one year later, she had improved somewhat according to maternal report. She was now attending a preschool program two days per week, and was doing well with her peers and loved going. The teacher saw no problems with her. Mrs. J., however, continued to see Annie as requiring a good deal of structure, as defiant, and as difficult to control. Mrs. J. complained particularly about Annie's uncooperative behavior when she was taken shopping, yet another example of inappropriate expectations. Sleeping and eating problems persisted and struggles in these areas were especially intense.

Thus problems around developmental tasks continued to be in evidence, clearly fueled by maternal difficulties conceptualizing her daughter's psychological needs or helping her to negotiate issues of separation-individuation and the establishment of autonomy and independence. It is particularly significant that Annie was able to separate successfully enough to attend a preschool program and that her teacher found her eager to play with other children and to participate in structured activities. Thus, with the emotional support provided by the preschool teacher, Annie was able to begin to reach out to others and to develop appropriately in certain areas.

When followed at age six, the marked discrepancy between Annie's behavior with her mother and with others was still apparent, and mother's perceptions of Annie still appeared to be quite distorted. During interactions with the project staff, Annie was a cooperative and sociable but a perceptibly needy youngster who appeared to be starved for attention, affection, and positive feedback. In school she was performing well, and her teacher saw no problems other than a tendency to talk too much to her neighbors. Classroom observations confirmed this report.

Unfortunately, Mrs. J. painted a very different picture. She complained of continued difficulties with discipline, inattention, overactivity, and poor impulse control. Indeed, Annie met DSM-III criteria for Attention Deficit Disorder with Hyperactivity according to maternal interview. In addition, Mrs. J. reported problems with bedwetting, nightmares, anxiety, fearfulness, and loneliness, portraying an anxious, unhappy, and possibly angry youngster, rather than a child with primarily externalizing problems. In contrast, Annie's teacher did not endorse one symptom on the Child Behavior Checklist!

Annie's problems would appear to stem primarily from insensitive parenting reflected in inappropriate expectations, negative perceptions, an inability to respond to and meet Annie's emotional needs, and difficulties setting appropriate limits. At age six, she remained a highly anxious youngster with low self-esteem and an overwhelming need to please. While it is obvious that Annie is not really a child showing Attention Deficit Disorder, she may well be at risk for continued problems with anxiety, sadness, and a poor self-concept.

DISCUSSION

The above cases illustrate two different patterns that we observed in the development of young children whose parents called us with complaints about problem behaviors. In cases such as David's, parents accurately reported truly difficult behaviors which endured as the children developed. These cases best exemplify early manifestations of the clinical syndrome of hyperactivity. In cases such as Annie's, parental perceptions were inconsistent with those of other observers and seemed to reflect an early parent-child mismatch and inappropriate parental expectations that in turn have led to continuing distress in the parent-child relationship. Although these children's actual behaviors were quite different, diagnostic procedures relying primarily on parental interview or questionnaire measures would result in labeling both children as hyperactive.

We observed other patterns that we do not have the space to describe in detail. For example, other children also exhibited high degrees of overactive, inattentive, aggressive, and noncompliant behaviors which persisted over time. In contrast to David, however, these children lived in families troubled by marital conflict, suggesting that social context effects also may have contributed to persistent problems. Even when parents were separated, conflict and dispute raged around these children, sometimes involving them directly, as parents battled for their allegiance. Where these conflicts have continued over time, it is impossible to disentangle the potential contributions of children's biological vulnerability from the effects of the anger, frustration, and emotional turmoil that are often responses to living in a disturbed family. In other words, if these same youngsters had lived in a more stable family, would their outcome have been better?

Other children were enrolled in our project by parents who complained of similar behaviors, but our observations suggested that their concerns revolved around less intense manifestations of these behaviors which are sometimes typical of preschoolers. Several of these children were little girls who exhibited "unladylike" behavior. After participating in our parent training groups, some of these parents appeared to understand more about the typical behavior of preschoolers and felt that they were managing their children more effectively. At follow-up, many of these parents reported improvements in their children's behavior and better relationships with them. These improvements were probably a function of developmental changes in the children as well as changes in parental attitudes and expectations.

Finally, the development of some children did not seem to fit into any of these patterns. For example, one of our most difficult preschoolers literally swung from the chandelier and was extremely active, irritable, and noncompliant at home, in our playroom, and at preschool. He responded, however, to positive, supportive, firm childrearing, and he gained more self-control as he matured. Despite a turbulent early course, he was doing well by age 5 and was not considered hyperactive. Several other children no longer showed severe behavior problems by age 6, but seemed to be demonstrating early signs of learning difficulties paired with continued attentional problems.

Thus, our longitudinal study suggests that despite similar parental complaints about symptoms akin to hyperactivity and related discipline problems, children followed a number of different developmental paths. Some cases seem to reflect characteristics inherent in the child, which may or may not persist as the child develops. Children such as David, whose hyperactivity persists despite early intervention and a stable family environment, probably exemplify actual hyperactivity as manifested in the preschool period. Others were merely going through a difficult phase of development and showing exaggerations of age-appropriate behavior. The problems of other children appear to reflect intolerant parental attitudes and/or ineffective childrearing practices, while still other children may be responding primarily to stresses in the social environment. Persistence of difficult behaviors in these children may depend upon the persistence of the problematic parental behaviors or environmental stresses. An understanding of behavior problems, including hyperactivity, in the preschool period, therefore, requires a thorough assessment of the child, the parents, and the social context as well as continued follow-up. Predictions about outcome in the individual case may be difficult, but early intervention may help to prevent, alleviate, or eliminate problems at school entry.

REFERENCES

Campbell, S. B. (1985). Hyperactivity in preschoolers: Correlates and prognistic implications. *Clinical Psychology Review, 5,* 405-428.

Campbell, S. B., Breaux, A. M., Ewing, L. J., & Szumowski, E. K. (1985, April). *Family characteristics and child behavior as precursors of externalizing symptomatology at school entry*. Paper presented at the biennial meeting of the Society for Research in Child Development, Toronto, Ontario.

Campbell, S. B., Ewing, L. J., Breaux, A. M., & Szumowski, E. K. (1986). Parent-referred problem three-year-olds: Follow-up at school entry. *Journal of Child Psychology and Psychiatry, 27*, 473-488.

Campbell, S. B., Szumowski, E. K., Ewing, L. J., Gluck, D. S., & Breaux, A. M. (1982). A multidimensional assessment of parent-referred behavior problem toddlers. *Journal of Abnormal Child Psychology, 10*, 569-591.

Fischer, M., Rolf, J. E., Hasazi, J. E., & Cummings, L. (1984). Follow-up of a preschool epidemiological sample: Cross-age continuities and predictions of later adjustment with internalizing and externalizing dimensions of behavior. *Child Development, 55*, 137-150.

Richman, N., Stevenson, J., & Graham, P. J. (1982). *Preschool to school: A behavioural study*. New York: Academic Press.

What Are Hyperactive Children Like as Young Adults?

John R. Kramer, PhD

ABSTRACT. In this article, the author describes two males with Attention Deficit Disorder (ADD) when they are referred as children to a clinic and when they are 21 years of age. The two case histories indicate that each subject behaves in a fairly consistent manner over time. Also, differences between these individuals at referral predict differences in their outcome as young adults: the subject who displays more aggression as a child engages in more antisocial activity as an adolescent and an adult. This finding is discussed in relationship to the literature.

Over the past 15 years, several follow-up studies have appeared in which individuals with Attention Deficit Disorders (ADD) are re-evaluated as young adults (e.g., Borland & Heckman, 1976; Feldman, Denhoff & Denhoff, 1979; Gittelman, Mannuzza, Shenker & Bonagura, 1985; Loney, Whaley-Klahn, Kosier & Conboy, 1983; Weiss, Hechtman, Milroy & Perlman, 1985). While these group studies provide valuable information about the outcome of ADD, it is useful to supplement statistics with case examples. In the present article, two young men who were formerly seen for ADD will be described as children and in their early twenties. Before presenting this information, the research project from which these subjects were drawn will be summarized.

Loney and her colleagues at the University of Iowa began investigating the outcome of hyperactive children in the early 1970s. The original sample consisted of 135 4- to 12-year-old boys who had been seen at the Child Psychiatry Outpatient Clinic for problems with overactivity, short attention span, impulsiveness, and/or learning problems. Seventy percent satisfied the DSM II criteria for

---

John R. Kramer is Principal Investigator of an ADD follow-up study in the Department of Psychiatry at the University of Iowa, Iowa City, IA 52242. This paper was supported by NIMH Grants MH-22659 and MH-41695.

© 1987 by The Haworth Press, Inc. All rights reserved.

Hyperkinetic Reaction of Childhood. All of these children were non-psychotic, had IQs over 70, and were free from obvious neurological or sensory dysfunction. Each boy began a trial of central nervous system stimulants, usually methylphenidate, within six weeks of referral.

Between 1973 and 1976, 124 of the 135 subjects returned for an investigation of adolescent outcome. A comprehensive set of data was collected from subjects, parents, and teachers (for a full description of the procedures, see Loney, Langhorne, Paternite, Whaley-Klahn, Blair-Broeker & Hacker, 1980). A second, young adult follow-up of this sample was begun in 1978 and is still in progress. Subjects are evaluated between the ages of 21 and 23 with a battery of tests, interviews, and questionnaires. Several major areas are assessed, including: (a) educational, marital, and vocational history; (b) intellectual and neuropsychological functioning; (c) behavior and personality; (d) psychiatric status; and (e) antisocial behavior and substance use.

A major goal of both the adolescent and adult follow-up studies has been to examine the outcome of individuals with ADD. A second purpose of these investigations has been to identify predictors of outcome. The reader is referred to Loney, Kramer, and Milich (1982) for detailed discussions of adolescent outcome; and to Loney et al. (1983), Weiss et al. (1985), and Gittelman et al. (1985) for descriptions of adult functioning. A brief summary of the main findings follows.

In most longitudinal studies, including the Iowa project, investigators have found a reduction of core ADD symptoms over time. Overactivity, impulsiveness and/or distractibility, however, usually persist to some extent. Other symptoms such as poor peer relationships, low self-esteem, and depression achieve prominence as the subject reaches adolescence. Among adult subjects, there is some suggestion that depressive disorders occur at elevated rates, although those data are drawn from young samples which are not yet beyond the age of risk for psychiatric disorders. A number of the subjects in our adult follow-up study are socially isolated and convey a clinical impression of sadness and lack of direction.

A common finding among ADD follow-up subjects is the presence of academic skill deficits. Most individuals in our project are at least several years behind in their reading, spelling, and/or mathematical skills, and some have a history of repeating grades. Few of our subjects have gone to college, and, as might be expected, the majority have worked at jobs which do not require much reading or writing.

Antisocial activity is perhaps the most controversial facet of outcome among ADD subjects. Debate centers around (a) whether ADD subjects are at increased risk for substance abuse and delinquent activity; and (b) what factors account for such behavior. Some researchers have obtained higher rates of substance use among ADD subjects than among controls (e.g., Gittelman et al., 1985), whereas other investigators find no differences (Kramer & Loney, 1982). Other forms of antisocial behavior are found more consistently to be elevated among ADD subjects (e.g., Gittelman et al., 1985; Loney et al., 1983).

Initially, researchers concluded that hyperactive children engaged in these behaviors because of frustration with school and peer rejection. Our data suggest a second explanation: ADD symptoms at referral do not correlate with later antisocial behavior; rather, it is aggression at referral (e.g., fighting and destruction of property) which predicts such problems. All of our subjects had ADD symptomatology. A number of them were also aggressive, and it is these latter individuals whom we have found to be at highest risk for later delinquent activity. Since many of the ADD subjects seen at clinics are also aggressive, it is not surprising that follow-up studies which are based on clinic cases obtain elevated rates of antisocial behavior.

METHOD

In order to illustrate this point, two subjects with different symptom patterns at referral were chosen from the original Iowa sample. The goal was to select one individual who exhibited primarily ADD symptoms and a second subject who exhibited ADD symptoms plus aggressive behavior. In order to accomplish this, a list was generated that displayed hyperactivity and aggression factor scores for each subject at referral. The scores were based on ratings of clinic reports. Several years ago, two judges had examined the reports and rated the degree to which each child displayed a variety of symptoms. A factor analysis of these symptom ratings at that time generated several factors. The first contained symptoms related to aggression (negative affect, aggressive interpersonal behavior, impulse control deficits); and the second contained symptoms indicative of hyperactivity (hyperactivity, inattention, judgement deficits).

Two subjects were selected on the basis of these dimensions; one had an above-average score only on the Hyperactivity factor (ADD Subject—"David"), and the other had above average scores on both the Hyperactivity and Aggression factors (ADD/Aggressive Subject—"Sam"). David had an IQ score of 96 at referral and 99 on two years prior to that; Sam had a prorated IQ of 117 at referral and a full IQ of 105 two years previously. On Hollingshead and Redlich's Two-Factor Index, where 1 represents highest Socioeconomic status and 5 represents lowest Socioeconomic status, David had a score of 4, and Sam had a score of 3.

After selecting the two cases, reports completed at referral by the clinic staff were used to generate a description of each subject as a child. The adult outcome of these individuals was compiled from selected interviews and tests administered at follow-up. These included a general interview that gathered information about education, employment, and dating/marriage; the Wechsler Adult Intelligence Scale (WAIS) and Wide Range Achievement Test (WRAT); the Kagan Matching Familiar Figures Test (MFF); a symptom checklist developed at Iowa (completed by subjects and by examiners); the SADS-L Diagnostic Interview; the National Survey on Drug Abuse; and a delinquency interview developed for this study.

RESULTS

ADD Subject: Referral

David was referred at age 9 to the clinic by his parents and family physician. Primary complaints included poor school performance, hyperactive behavior, and distractibility. David was noted to be an overly excitable child who teased peers and interfered with their activities; he consequently had a difficult time making friends. The subject's mother described him as having a "learning block." David had difficulty remembering colors, numbers, and letters. He could not easily follow verbal directions, even when they were repeated. David was also noted to stutter and talk too fast. His family consisted of two younger siblings; his mother, who was a housewife; and his father, a farmer.

In the clinic, the psychiatrist found that " . . . as the interview progressed, he became more hyperactive, rocking back and forth in the chair, and becoming loud, talkative, and boisterous. There were certain occasional lags in his attention.'' The outpatient diagnosis was Hyperkinetic Reaction of Childhood and Specific Learning Disturbance. Recommended treatment included methylphenidate and imipramine (for bedwetting) and outpatient therapy with David and his parents. The family agreed to this arrangement and attended the clinic for three years; during that period David remained on medication, although his methylphenidate was changed to dextroamphetamine and his imipramine to nortriptyline.

ADD Subject: Follow-Up

At age 21, David was living with his parents and dating one woman steadily. He had graduated from high school and completed training in auto mechanics. David was unemployed at the time of follow-up and had never worked full-time. His Full Scale WAIS IQ was 102; grade level and percentile scores on the WRAT were: Word Recognition, 6.3 (9th); Spelling, 6.2 (13th); and Arithmetic, 8.8 (63rd). David made six errors on the Kagan MFF with an average latency of 12.8 seconds per item; compared to norms, both scores suggested impulsiveness.

The examiners found David to be somewhat anxious and occasionally grumpy. He was noticeably fidgety and shifted in his seat. At times, the subject was inattentive. David endorsed six checklist items to describe himself: "on the go," "overactive," "restless," "fiddles with objects," "excitable," and "impulsive." He indicated during the SADS-L Interview that he had experienced major depression at age 19 when his girlfriend left him.

David did not use drugs or alcohol to any great extent. He estimated that he had consumed alcohol on 4 days of the previous month. David first tried marijuana within the past year and had used it less than ten times; he stated that he did not plan to smoke it again. David denied trying any other substances and claimed he did not intend to in the future.

The subject's antisocial history was moderate. David broke into a gas station at age 17 but claimed he did nothing once inside. He was placed on probation for one year as a result of this incident. Other police contacts included (a) a warning because he was driving with opened beer bottles and suspected of OMVI; and (b) a suspended license at age 19 because of repeated speeding. The subject had two physical fights in the past year. No other major personal or property offenses were recalled.

ADD/Aggressive Subject: Referral

Sam was referred to the University clinic by his local physician at age 11. The school complained of distractibility and restless behavior which sometimes took the form of clowning in class. Despite Sam's attempts to gain approval in this way, peer relationships were not good. This was due in part to his obesity, impulsiveness, and bad temper. Sam was a natural target for teasing, and he often reacted explosively. During one conflict, he seized his father's shotgun and threatened a friend with it. Sam's mother noted that he had been a temperamental and stubborn infant and continued to be oppositional. Sam was destructive with toys and often lied to his parents. The family consisted of Sam's father, a used car salesman; his mother, a housewife; and a younger sister.

The clinic staff noted that Sam fidgeted during the evaluation. He spoke in a rushed fashion and related to the examiners in an overly demonstrative manner. Poor performances on the Bender-Gestalt and on the Block Design subtest of the WISC suggested the possibility of significant visual-motor problems. Sam was diagnosed as Hyperkinetic Reaction of Childhood. It was recommended that a trial of dextroamphetamine be instituted in conjunction with outpatient therapy for Sam and his parents.

ADD/Aggressive Subject: Follow-Up

After a few years of periodic counseling and dexedrine treatment, Sam was admitted at age 15 to the inpatient ward in Child Psychiatry for two weeks because of exhibitionism and sex play with children, which on one occasion involved an attempt at intercourse with his 9-year-old sister. Sam behaved well during his two-week stay on the ward; one psychiatrist noted that Sam was either a " . . . charming fellow . . . or we are being had."

When the subject was re-evaluated at age 21, he was renting an apartment and did not date. Sam was unemployed at follow-up; over the previous few years, he had worked at 12 manual labor jobs, including delivery boy, car washer, and trash collector. Sam graduated from high school but received no further education. His WAIS Full Scale IQ was 104; on the WRAT, he recognized words at Grade Level 8.7 (47th percentile), spelled at 8.9 (50th percentile), and solved arithmetic problems at 5.9 (19th percentile). On the

Kagan MFF, Sam performed in an impulsive manner by making ten errors and responding to items in an average of 7.3 seconds. During follow-up testing, Sam was noted by the examiners to be overly friendly and boastful. He sometimes laughed when describing his problem behaviors. The subject responded to questions impulsively; he often interrupted the examiner to insert a word or make a funny remark. Sam also appeared restless and fidgeted during the evaluation. When he was asked to describe himself on a checklist, Sam endorsed the items: "always on the go," "can't sit still," "forgetful," "messy," "impatient," and "disobeys rules." On the SADS-L Interview, he satisfied the criteria for phobic reactions to heights and darkness.

Sam used substances to a greater extent than David. He estimated that he had consumed alcohol on 11 to 20 days of the past month. The subject stated that over the past year he had driven about 50 times while drinking and an equal number of times while intoxicated. Sam began using marijuana several years prior to follow-up. He estimated that he had used it on 15 days out of the past month, and he planned to continue smoking. Sam first tried hashish at age 19, had used it during the previous month, and planned to use it again. The subject also admitted to experimenting with prescription stimulants and tranquilizers over the past year. Although he had not used cocaine, hallucinogens, heroin, or methadone, Sam stated that he might try them at some future point. He also indicated that he had helped set up a bust of PCP users.

The subject's most severe legal sanction was a suspended license for one year because of multiple violations; just before he lost his license, Sam was caught knocking over signs with his car. The subject admitted that he frequently sped, made illegal turns, and ran stop signs. Sam stole auto parts from work several times within the previous year. He recalled spray painting signs at age 18 and stealing portable toilets from a homecoming celebration. Sam also admitted that he did not always repay debts, and he remembered at least one period during which he wandered about with no permanent residence.

The subject did not meet SADS-L interview criteria for Antisocial Personality for three reasons: (a) he denied that he fought frequently; (b) he stated that he had not been arrested for serious offenses; and (c) he claimed that he had close interpersonal relationships. The SADS-L criteria for this disorder are stringent, and one might conclude from a less structured evaluation that the subject did display Antisocial Personality.

DISCUSSION

If one examines the longitudinal data from these two individuals, it is possible to identify continuities of behavior which might not be discernible in group statistics. When David and Sam were evaluated at age 21, the examiners did not have access to their earlier records. Nevertheless, descriptions of these subjects as adults bear a strong resemblence to their clinic reports from childhood.

David originally came to the clinic because of problems with hyperactivity and distractibility. At follow-up, he was still restless and fidgety; the examiners also detected continuing problems with attention. The clinic psychiatrist's description of David cited earlier indicates that he was an excitable child. At the adult follow-up 12 years later, David characterized himself as still "excitable." David's "learning block" continued into his adult years. Although his Full Scale, Verbal, and Performance IQ scores at age 21 were close to 100, his reading and spelling skills on the WRAT were considerably below average. David's drawings on the Bender-Gestalt at follow-up contained a moderately high number of errors, suggesting that he continued to have difficulties with certain visual-motor tasks.

As a child, Sam was described by his parents as someone who, when he " . . . wants to demonstrate love . . . becomes overly affectionate." Ten years later, our examiner described him with the checklist item "overly friendly." Sam behaved impulsively at referral, and he continued to act this way during the follow-up evaluation by interrupting the examiner and acting silly. Examiners who tested Sam as a child noted that he fidgeted and had trouble sitting still; a similar set of observations was made by the follow-up examiners. Sam described himself at 21 as someone who disobeys rules, which is consistent with his mother's earlier descriptions of oppositional behavior. Sam was described at referral as "literally afraid of everything," including the dark; this particular fear persisted into his adult years.

The case history data illustrate not only within-subject consistencies, but also between-subject differences. Sam's follow-up interview contains a greater number of norm-violating behaviors than does David's, a finding which is consistent with research on the prognostic significance of early aggression. Sam's sexual behavior during adolescence was sufficiently aggressive to warrant inpatient treatment. His involvement with drugs began earlier and was more extensive. Sam also drank more frequently and drove repeatedly while intoxicated. Both subjects had their driver's license suspended; David's suspension, however, resulted from speeding, whereas Sam's suspension was based on more destructive behaviors, such as driving over signs. Sam committed several acts of vandalism and stole material from work; although David did break and enter, he denied that he otherwise destroyed property or had ever engaged in theft.

Psychiatric and vocational outcome do not discriminate as clearly between the two subjects as does antisocial behavior. David and Sam each satisfied SADS-L criteria for one psychiatric disorder, and it is unclear if a week-long episode of major depression constitutes a better or worse psychiatric outcome than strong fears of darkness and heights. Both subjects have poor job histories; David never worked full-time, and Sam held jobs only briefly. Since the subjects were evaluated only part-way through the risk periods for psychiatric disorders and vocational problems, it is understandable why major differences in these two areas were not found.

Differences in behavioral outcome that did emerge between the two subjects were generally consistent with Sam's higher aggression score at referral. Sam's referral IQ score was also higher than David's, but this does not account for the majority of differences between the two subjects' antisocial behavior; in general (e.g., Loney et al., 1983), higher referral IQ is associated with a lower rather than a higher rate of subsequent norm-violating behavior.

The reason why aggression in ADD children predicts a worse outcome is not yet completely understood. One explanation may be that aggressive children are more likely to have pathological families. During the follow-up evaluation, both subjects were asked a set of questions which probed their parents' legal and drug histories. Sam's father met all criteria for alcoholism and most criteria for antisocial personality; in contrast, neither of David's parents did. Sam's father may have contributed to his son's aggression during childhood and to his antisocial behavior during adolescence. This contribution could have been environmental and/or genetic in nature.

Sam and David's histories reveal both continuities and differences. Each of them displayed particular behaviors and traits consistently over a ten- to twelve-year time period. The two individuals also exhibited differences at age 21 which were predictable from differences between them at referral. These cases illustrate the association between aggression in childhood and antisocial activity in early adulthood. It remains to be determined whether this association will persist through later stages in life.

REFERENCES

Borland, B. L., & Heckman, H. K. (1976). Hyperactive boys and their brothers: A 25-year follow-up study. *Archives of General Psychiatry, 33*, 669-675.

Feldman, S. A., Denhoff, E., & Denhoff, J. I. (1979). The attention disorders and related syndromes: Outcome in adolescence and young adult life. In E. Denhoff & L. Stern (Eds.), *Minimal brain dysfunction: A developmental approach* (pp. 133-148). New York: Masson.

Gittelman, R., Mannuzza, S., Shenker, R., & Bonagura, N. (1985). Hyperactive boys almost grown up: I. Psychiatric status. *Archives of General Psychiatry, 1985, 42*, 937-947.

Kramer, J. R., & Loney, J. (1982). Childhood hyperactivity and substance abuse: A review of the literature. In K. Gadow & I. Bialer (Eds.), *Advances in learning and behavioral disabilities* (Vol. 1, pp. 225-259). Greenwich, CT: JAI Press.

Loney, J., Langhorne, J. E., Paternite, C. E., Whaley-Klahn, M. A., Blair-Broeker, C. T., & Hacker, M. (1980). The Iowa HABIT: Hyperkinetic/aggressive boys in treatment. In S. Sells, R. Crandall, M. Roff, J. Strauss, & W. Pollin (Eds.), *Human functioning in longitudinal perspective: Studies of normal and psychopathic populations* (pp. 119-140). Baltimore: Williams & Wilkins.

Loney, J., Kramer, J. R., & Milich, R. (1982). The hyperactive child grows up: Predictors of symptoms, delinquency, and achievement at follow-up. In K. Gadow & J. Loney (Eds.), *Psychosocial aspects of drug treatment for hyperactivity* (pp. 381-415). Boulder: Westview Press.

Loney, J., Whaley-Klahn, M. A., Kosier, T., & Conboy, J. (1983). Hyperactive boys and their brothers at 21: Predictors of aggressive and antisocial outcomes. In K. Van Dusen & S. Mednick (Eds.), *Prospective studies of crime and delinquency* (pp. 181-207). Boston: Kluwer-Nyhoff.

Weiss, G., Hechtman, L., Milroy, & Perlman, T. (1985). Psychiatric status of hyperactives as adults: A controlled prospective 15-year follow-up of 63 hyperactive children. *Journal of the American Academy of Child Psychiatry, 24*, 211-220.

TREATMENT

What Do We Know About the Use and Effects of CNS Stimulants in the Treatment of ADD?

William E. Pelham, Jr., PhD

ABSTRACT. The most common form of intervention for ADD children is medication with a central nervous system stimulant. This chapter reviews what is known about the use and effects of psychostimulant drugs in the treatment of ADD. Approximately 70% of the ADD children treated with stimulants demonstrate short-term improvement on most measures of behavior and academic performance in classroom settings, for which the stimulants are typically prescribed. However, stimulants as a sole form of intervention have not resulted in an altered long-term prognosis for treated children. Suggestions for clinical use of stimulants are offered.

Pharmacological intervention with one of the central nervous system (CNS) stimulant drugs has been the most common mode of treatment for ADD for the past 25 years. Although other classes of drugs have been evaluated for ADD, they have not been widely used, and they have not met with much success. In contrast, 80% to 90% of all ADD children have been treated with a stimulant at some point in time, and the drugs have been shown to be quite effective in the short-term management of the disorder. The purpose of this chapter is to describe the effects that stimulant drugs have on ADD children and their role in a comprehensive treatment for ADD. An

---

William E. Pelham, Jr., is Associate Professor of Psychiatry, Department of Psychiatry; Director of the Attention Deficit Disorders Program of the Western Psychiatric Institute and Clinic, University of Pittsburgh, 3811 O'Hara Street, Pittsburgh, PA 15213.

© 1987 by The Haworth Press, Inc. All rights reserved.

attempt will be made to blend basic information about the drugs with practical information that will be useful to professionals faced with the question of whether and/or how to medicate an ADD child.

PHARMACOLOGY OF THE CNS STIMULANTS

The three psychostimulants that are commonly employed with ADD children are methylphenidate (MPH; Ritalin), dextroamphetamine (Dexedrine), and pemoline (Cylert), with many more children receiving MPH than dextroamphetamine or pemoline. These drugs have relatively brief half lives; they take effect quickly and their effects wear off quickly. Behavioral effects are observed within 30 minutes (regular MPH and dextroamphetamine) to 90 minutes (pemoline) following ingestion. The behavioral effects of MPH and dextroamphetamine peak approximately 2 hours post ingestion and generally disappear 2 to 3 hours later. Pemoline has a later peak and its effects last for a total of 8 to 10 hours. Both MPH and dextroamphetamine are available in timed release forms, the effects of which also last 8 to 10 hours. Pemoline and the timed release versions of MPH and dextroamphetamine are thus administered once daily in the morning, while regular MPH and dextroamphetamine are typically given with both breakfast and lunch. There is considerable controversy regarding what dose is best (see the following discussion). Generally, MPH is administered twice daily in doses that range from .25 to .75 mg/kg of body weight, with dextroamphetamine doses half this amount. Pemoline dosing levels are not as well determined as for the other stimulants, but it appears that doses 4 to 6 times those of a single dose of MPH are required to yield the same behavioral effect.

The precise neural mechanisms of the stimulants' effects in humans are as yet unknown (Solanto, 1984). They are presumed to enhance broadly the effects of CNS neurotransmitters (the catecholamines—dopamine and norepinephrine) by increasing their availability at the synapse. Whether the various stimulants share the same neurochemical mechanism is also unknown. Some children respond positively to one stimulant and not to the other two, implying that the drugs may have different neurochemical mechanisms that interact with individual differences in brain biochemistry.

Although at one time, stimulant effects on ADD children were thought to be different from those that would be obtained with normal children (thus the old notion of the paradoxical effect), it is now clear that stimulants affect normal children and adults in the same manner that they affect ADD children (Rapoport, Buchsbaum, Weingartner, Zahn, Ludlow & Mikkelsen, 1980). Specifically, on a variety of measures, the drugs increase activity or arousal in the CNS. How these changes in CNS activity are related to changes in ADD children's behavior is not well understood.

EFFECTS OF CNS STIMULANTS

Whatever the underlying physiological action of the stimulants, a great deal is known about their effects on cognitive and behavioral measures. On laboratory measures of sustained attention and impulsivity, two core cognitive deficits of ADD, stimulants improve ADD children's performance. On a typical sustained attention or vigilance task, relative to a placebo condition, medicated children make fewer impulsive responses to nontarget stimuli, and they maintain attention and miss fewer targets. These effects are most pronounced during the later portion of the task. The beneficial medication effects are observed on concurrent electrophysiological measures of brain activity.

Similarly positive effects of stimulants have been reported on a wide variety of laboratory tasks of information processing and learning (for reviews see Cantwell & Carlson, 1978; Conners & Rothschild, 1968; Pelham, 1986; Swanson & Kinsbourne, 1979). For example, beneficial stimulant effects (approximately 25% average improvement) have been consistently reported on paired-associate learning tasks and nonsense word spelling tasks.

Given the results from laboratory tasks, it should not be surprising that beneficial stimulant effects have been observed in ADD children's classroom performance. Stimulant-induced reductions in impulsivity and increases in sustained attention manifest themselves most clearly in the classroom as decreases in classroom disruptiveness and increases in on-task behavior. Relative to unmedicated ADD children, those receiving stimulants are less likely (a) to talk out inappropriately in class, (b) to bother peers who are working, (c) to violate classroom rules or engage in other behaviors that require teacher attention, and (d) to interact aggressively and otherwise inappropriately with peers. These changes have been reported in a very large number of studies and are clearly evident whether children are observed directly or rated by their teachers. Instead of engaging in these behaviors, medicated children spend more time behaving appropriately, such as being on task in the classroom and completing more of their assigned academic work, often with increased accuracy. Studies in laboratory analogue settings have shown that stimulant drugs affect ADD children's behavior during structured parent-child interactions in the same way that the drugs affect the children's behavior at school. Medicated children are less disruptive and more compliant than nonmedicated children (e.g., Barkley & Cunningham, 1979). Because medication is rarely given in the evenings, however, these potentially beneficial effects on parent-child interactions are rarely realized in practice. The effects of a morning or noon dose of stimulant have usually worn off by evening. Parents usually will not see drug-related behavior changes and thus cannot be relied upon to detect them.

Unfortunately, not all ADD children show positive responses to stimulant drugs. Approximately 70% of ADD children respond positively to a stimulant regimen, but the remaining 30% show either an adverse response or none. Of those who do respond, only a minority show sufficient improvement for their behavior to fall within the normal range; the rest are improved but their behavior is not normalized. For example, the average rating on the ACTRS (see Conners in this issue) for a group of unmedicated ADD children is often 20. In many drug studies, this mean rating decreases to 10, a 50% reduction. This compares with a score of only 2 or 3 that nondeviant children receive on this scale. For the majority of positive medication responders, then, treatment in addition to medication is required to bring them within a normal range of functioning.

One variable on which the degree of a child's response to a stimulant drug depends is the administered dose. Generally, the larger the dose, the larger the drug effect, and the dose-response relationship is usually linear. An example of this pattern can be found in Figure 1 from Pelham, Bender, Caddell, Booth, and Moorer (1985). Some authors have reported, however, dose-response curves that vary over tasks and are not linear. For example, Sprague and Sleator (1977) presented data showing that maximal improvement on a cognitive, laboratory measure was obtained with .3 mg/kg of MPH, whereas maximal improvement in teacher ratings on the ACTRS was not apparent until the dose was 1.0 mg/kg, at which level children's performance on the cognitive task was impaired relative to placebo. Although most professionals agree with Sprague and Sleator's conclusion that classroom disruptiveness may require higher doses of stimulants to be corrected than cognitive deficits re-

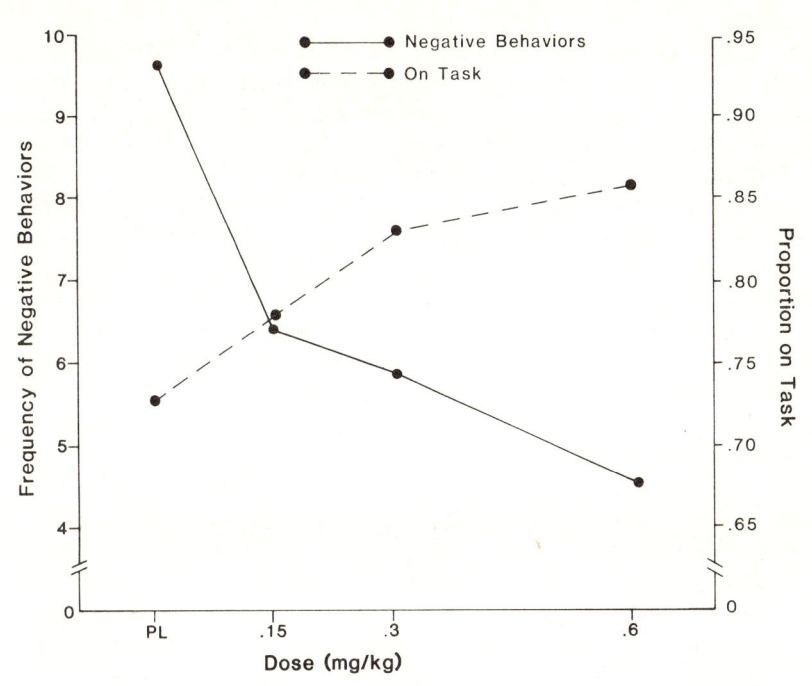

FIGURE 1. Observed negative behaviors and on-task behavior as a function of methylphenidate dose. *NOTE*: From "The dose-response effects of methylphenidate on classroom academic and social behavior in children with attention deficit disorder" by W. E. Pelham, M. E. Bender, J. Caddell, S. Booth & S. H. Moorer, 1985, *Archives of General Psychiatry*, *42*, p. 948-952. Copyright 1985 by the American Medical Association. Reprinted by permission.

quire, the precise point at which stimulants have adverse effects on cognition is not yet determined. Numerous studies that have administered MPH doses up to .75 mg/kg have found improvement on cognitive tasks. It needs to be noted that there are large individual differences in response to stimulants. Group data do not adequately reflect dose effects for the individuals who make up the group, and a standard dose will not be appropriate for every child.

In addition to potential adverse stimulant effects on cognition, other treatment emergent symptoms (TES) are sometimes associated with stimulant treatment. For example, long-term stimulant treatment (2 to 4 years) with doses of 40 mg of MPH per day results in a reduction in the rate of weight gain and to a lesser extent height gain in treated children (Mattes & Gittelman, 1983). This effect is dependent on the total cumulative dose that the child takes, however, and can be avoided by keeping the dose low and by having the child not take medication on weekends and holidays. Although not known for certain, this reduction in growth may result from the anorectic effect of stimulants. Medicated children usually have a reduced appetite and eat less at meals compared to placebo days. Stimulants can also cause socially withdrawn behavior in treated children (Pelham & Hoza, in press). Finally, stimulants have been reported to exacerbate motor tics when present in treated children, with some professionals even suggesting that stimulants can precipitate Tourette's syndrome. Although many of the most serious TES are rare, their potential impact means that medicated children should be very carefully monitored to insure that tics, social withdrawal, cognitive impairment, etc., are not caused or exacerbated by stimulant therapy. Several standard rating scales for evaluation of TES are available (Rapoport, Conners & Reatig, 1985).

CLINICAL ASSESSMENT OF STIMULANT EFFECTS

The question facing every practitioner is how to decide whether a stimulant should be used with a child and what the correct dose should be. The "whether" decision is rarely made systematically in practice. Most often, parents contact their child's physician at the insistence of the school, and the physician prescribes medication (or does not) based on his or her knowledge and beliefs regarding the role of stimulants in treatment of ADD. Only rarely does the physician obtain information systematically about the child from the setting that initiated the referral—the school. A starting dose is selected and increases are made based on parental reports until optimal drug response is presumably obtained and no adverse side effects are apparent. Again, lack of contact with the school is the norm (Gadow, 1981). This procedure, which is fairly standardly recommended in pediatric psychopharmacology, has numerous shortcomings. It fails to gather systematically the information that should form the basis of a decision regarding whether to medicate (i.e., whether the medication has beneficial effects on the symptoms that prompted the referral); it confounds increasing medication dose with events that may be changing over time; and it results in children receiving the highest possible dose of stimulant.

Alternative procedures for determining the utility of stimulant therapy for a child and for selecting the proper drug dose have been advocated in recent years. Because of their short half lives, stimulants can be alternated frequently with placebo for brief periods of time (days or weeks) in an assessment that may take a total of 2 to 5 weeks, depending on the number of drugs or doses evaluated. All of the investigators currently conducting and writing about systematic medication assessments employ this alternating drug-condition protocol as well as placebo-controlled, double-blind procedures. All use the assessment to make data-based recommendations concerning the utility of stimulant medication for a target child's treatment. Different teams, however, use different measures in their assessments. For example, Swanson and his associates (e.g., Swanson & Kinsbourne, 1979) employ a laboratory test of paired associate learning as the primary measure of stimulant effect. Other researchers use clinic observations of parent-child interactions, laboratory vigilance tasks, and teacher ratings. In addition, Rapport, Stoner, DuPaul, Birmingham, and Tucker (1985) gather classroom observations of on-task behavior. These clinical researchers all work from outpatient clinics, but similar assessments have been conducted in inpatient settings (e.g., Wells, Conners, Imber & Delameter, 1981).

As might be inferred from the variety of measures that have been employed, there is disagreement among professionals regarding the best measure of medication effects. Given this disagreement, multimethod assessments have been advocated. For example, the assessment conducted in my laboratory employs a large number of measures in order to gain as comprehensive a picture of stimulant effects as possible on each child. The assessment is conducted on children participating in a summer day treatment program (STP). Using a double-blind, placebo-controlled protocol in which active drug and placebo are alternated daily, we have assessed more than 150 children over the past few years (Pelham & Hoza, in press). As others' our purpose is to determine whether stimulants have a beneficial effect on a child and how they should be used in the child's long-term treatment.

The results of an assessment on one child, Matt, from our 1985 program are presented in Table 1. Matt was a severely hyperactive, 8-year-old child who also met diagnostic criteria for conduct disorder and learning disability. He had a long history of behavior and learning problems in school, had been retained once, and had been referred to the summer treatment program by his school psychologist. Table 1 gives an excellent portrayal of the effects of .3 and .6 mg/kg MPH b.i.d. on Matt's behavior and performance in a wide variety of domains, and it illustrates many of the points we noted above regarding stimulant effects on ADD children.

Table 1

Matt's Medication Assessment Results

Variable Measured	Placebo (7 days) Avg.	SD	.3 mg/kg MPH b.i.d. (6 days) Avg.	SD	% Change from PL	.6 mg/kg MPH b.i.d. (6 days) Avg.	SD	% Change from PL
Daily Frequencies:								
Following Rules	14.85	4.75	18.32	2.45	23.42	19.65	1.47	32.39
Noncompliance	12.71	9.39	7.33	6.76	-42.33	5.83	2.71	-54.15
Positive Peer Behaviors	35.5	9.92	70.59	21.9	98.84	87	31.34	145.06
Conduct Problems	.83	.75	.2	.43	-75.91	.15	.41	-80.73
Negative Verbalizations	8.16	5.27	6.8	3.11	-16.68	3.83	2.48	-53.07
Time Outs:								
Number per day	1.7	1.37	.15	.41	-90.66	.32	.51	-80.72
Minutes per day	38	45.31	1.66	4.08	-95.64	2.66	4.55	-93.01
Average minutes per Time Out	16.93	11.85	10	0	-40.97	8	4.24	-52.79
Classroom:								
% On Task	80	23.09	94.28	9.82	17.85	94.28	9.82	17.85
% Following Rules	57.14	20.89	70.56	36.64	23.5	96.13	5.26	68.25
Timed Math:								
Number Attempted	11.42	2.96	12.42	3.44	8.75	14.57	3.41	27.57
Percentage Correct	76	36.57	91.13	11.21	19.92	80	27.64	5.25
Timed Reading:								
Number Attempted	14	6.08	18.5	11.64	32.14	15.28	7.66	9.14
Percentage Correct	63.84	24.28	78.5	9.07	22.93	78	28.9	22.15
Seatwork:								
% Completion	56	41	81	35	44.64	87	16	57.14
% Correct	75	16	70	22	-5.34	83	15	10.66
Nonsense Spelling								
Number of Errors	70	0	46	0	-34	-	-	-
Teacher Rating (Conners Short Form)	16.5	.69	5	6.24	-69.7	4	0	-75.76
Counselor Rating	179.16	24.84	101.33	0	-43.46	99.33	14.73	-44.57
Pos. Daily Report Card (% of Days Received)	50	54	100	0	100	100	0	100
Observed Interactions:								
Positive Peer	94.4	5.01	98.8	1.64	4.66	97	3.45	2.75
Negative Peer	5.18	5.46	.4	.88	-92.31	2.25	2.87	-56.66
No Interactions	.4	.88	.8	1.29	100	.75	1.5	87.5

For example, on the daily frequency counts of how often Matt followed the rules as well as the measure of following rules in the classroom, MPH resulted in substantial improvement. Further, Matt was on task more in the classroom while medicated, and he completed more of his seatwork and performed more math and reading problems in timed tasks. When medicated, his accuracy on both the reading and the math tasks improved, and he made fewer errors in a nonsense spelling learning task. As the table indicates, his teacher's and his counselors' ratings of his behavior improved dramatically on medication days. When medicated, Matt was also more compliant with adult requests, as indicated by the 50% decrease in recorded noncompliance.

Stimulant-induced changes in Matt's antisocial behaviors are shown by the decreases in daily occurrences of conduct problems (stealing, lying, aggression, and destruction of property), negative verbalizations (e.g., cursing, name-calling, talking back), and observed negative peer interactions on the playground. It is interesting to note that these decreases were accompanied by increases both in daily frequency counts of positive peer behaviors (e.g., helping, sharing, saying something nice) and in observed positive peer interactions on the playground.

The data presented in Table 1 show that there were many intraindividual differences in the shapes of Matt's dose-response curves on different measures. Although his response to the two doses of MPH continued to improve in a linear manner as dose was increased on some measures (e.g., positive peer behaviors, classroom following rules), maximal response was reached on other variables at the low dose. Sometimes when maximal response was reached at .3 mg/kg, the higher dose yielded only slight incremental improvement (e.g., noncompliance, teacher rating), whereas the higher dose appeared to worsen behavior on other measures (e.g., percentage correct on timed math, observed negative peer interactions). Finally, on several measures Matt showed maximal improvement only when the high dose was used (e.g., negative verbalizations).

As do all other laboratories in which medication is assessed, we have counselors, teachers, and parents complete checklists designed to pick up any TES that would contraindicate a medication recommendation. Matt showed no TES on .3 mg/kg days, but he exhibited pronounced irritability and anorexia on days on which he received .6 mg/kg MPH.

It is noteworthy that the basic treatment in the STP was a highly structured behavior modification program that included a point system, response cost, time out, and a variety of other components. The medication assessments are conducted while the behavioral intervention is in effect. Matt's behavior on placebo days was quite deviant and thus reflected deficits that were not corrected with as intensive and comprehensive a behavioral intervention as we could design. Matt appeared to need both medication and a behavior modification program in order to show maximal improvement in his behavior.

Given the TES observed on the higher dose and the clearly beneficial effects of .3 mg/kg MPH on many of the symptoms for which Matt had been referred, we recommended that Matt receive that dose of MPH b.i.d. in conjunction with a behavioral intervention in his regular classroom setting. For the past year, he has been receiving that treatment regimen as well as resource room help for his learning problem, and his mother has continued to receive parent training. Matt's behavior in school is much improved, compared to the previous year when he was not medicated.

LONG-TERM EFFECTS OF STIMULANT TREATMENT

Despite the often dramatic salutary short-term effects of stimulants, there is as yet no evidence that long-term stimulant therapy modifies the dismal adolescent and adult prognoses of ADD children. There has been considerable debate regarding why these short-term benefits fail to translate into long-term changes. Excessive doses, lack of compliance with prescribed regimens, absence of an effect on critical behavioral domains, lack of necessary concurrent interventions, and presence of a persistent, underlying biological deficit are all reasons that have been postulated to account for this discrepancy, but none have been systematically evaluated in outcome studies. Some authors have argued that the discrepancy is only apparent, and that the long-term studies have failed to show long-term benefits because the studies are seriously flawed.

No resolution of this paradox is currently evident. The hope is that the effectiveness of long-term stimulant therapy will be increased if children undergo a systematic, clinical assessment as described above prior to beginning a stimulant regimen and if adequate monitoring procedures are used to follow a medicated child. For example, most professionals recommend using a placebo probe at yearly intervals to insure that a child continues to show a positive response to a stimulant. Further, parents and school personnel need to be consistent in their administration of medication. Finally, there is widespread agreement that stimulants should not be used in isolation in the treatment of ADD. A number of studies have demonstrated quite clearly that the combination of behavioral and stimulant interventions are more effective in the short-term than either treatment alone (Pelham & Murphy, 1986). As it was for Matt in our case example, the combination of these two modalities is becoming the treatment of choice for ADD.

The importance of concurrent intervention cannot be overemphasized. All too often an ADD child's dramatic, short-term improvement with medication removes the pressure from parents and teachers to invest the necessary effort in concurrent, comprehensive treatments, such as behavioral interventions. Parents then rely on the medication alone to solve the problem. Mental health professionals, educators, and physicians need to be especially alert to this possibility and to take action to prevent it from happening.

A final question that merits attention is how long a child should receive stimulant therapy. One approach has been to use stimulants for limited time periods (e.g., 3 to 6 months) to stabilize a problematic situation and to facilitate the initiation of other treatments. Although this approach has been frequently advocated, there are no guidelines suggesting how it is to be done, and no data that conclusively demonstrate its effectiveness.

Maintenance stimulant regimens are more common. There are no guidelines, however, that show how long a maintenance regimen should be continued, assuming a child continues to show a good response. Several years ago, the zeitgeist was that stimulant therapy needed to be terminated when the target child reached puberty, as the medication would lose its effectiveness after that point. Recent studies have suggested, however, that stimulants have beneficial effects both on ADD adolescents (Varley, 1983) and on some adults who have residual ADD (Wender, Reimherr, Wood & Ward, 1985). At present, then, there are no data that provide a definitive answer to the question of how long a child should receive a stimulant drug. Based on current information, the answer appears to be "as long as periodic, systematic evaluations reveal a therapeutic effect for the child."

REFERENCES

Barkley, R. A., & Cunningham, C. E. (1979). The effects of methylphenidate on the mother-child interactions of hyperactive children. *Archives of General Psychiatry, 36*, 201-208.

Cantwell, D. P., & Carlson, G. (1978). Stimulants. In J. S. Werry (Ed.), *Pediatric psychopharmacology* (pp. 171-207). New York: Brunner/Mazel.

Conners, C. K., & Rothschild, G. H. (1968). Drugs and learning in children. In J. Hellmuth (Ed.), *Learning disorders* (Vol. 3, pp. 191-217). Seattle, WA: Special Child Publications.

Gadow, K. D. (1981). Drug therapy for hyperactivity: Treatment procedures in natural settings. In K. D. Gadow & J. Loney (Eds.), *Psychosocial aspects of drug treatment for hyperactivity* (pp. 325-378). Boulder, CO: Westview Press.

Mattes, J. M., & Gittelman, R. (1983). Growth of hyperactive children on maintenance regimens of methylphenidate. *Archives of General Psychiatry, 40*, 317-321.

Pelham, W. E. (1986). The effects of stimulant drugs on learning and achievement in hyperactive and learning-disabled children. In J. K. Torgeson & B. Wong (Eds.), *Psychological and educational perspectives on learning disabilities* (pp. 259-296). New York: Academic Press.

Pelham, W. E., Bender, M. E., Caddell, J., Booth, S., & Moorer, S. H. (1985). The dose-response effects of methylphenidate on classroom academic and social behavior in children with attention deficit disorder. *Archives of General Psychiatry, 42*, 948-952.

Pelham, W. E., & Hoza, J. (in press). Behavioral assessment of psychostimulant effects on ADD children in a summer day treatment program. In R. Prinz (Ed.), *Advances in behavioral assessment of children and families* (Vol. 3). Greenwich, CT: JAI Press.

Pelham, W. E., & Murphy, H. A. (1986). Attention deficit and conduct disorders. In M. Hersen (Ed.), *Pharmacological and behavioral treatments: An integrative approach* (pp. 108-148). New York: Wiley.

Rapoport, J. L., Buchsbaum, M. S., Weingartner, H., Zahn, T. P., Ludlow, C., & Mikkelsen, E. J. (1980). Dextroamphetamine: Cognitive and behavioral effects in normal and hyperactive boys and normal men. *Archives of General Psychiatry, 37*, 933-943.

Rapoport, J., Conners, K. C., & Reatig, N. (Eds.). (1985). Rating scales and assessment instruments for use in pediatric psychopharmacology research. *Psychopharmacology Bulletin, 21*, 713-1125.

Rapport, M. D., Stoner, G., DuPaul, G. J., Birmingham, B. K., & Tucker, S. (1985). Methylphenidate in hyperactive children: Differential effects of dose on academic, learning, and social behavior. *Journal of Abnormal Child Psychology, 13*, 227-244.

Solanto, M. V. (1984). Neuropharmacological basis of stimulant drug action in attention deficit disorder with hyperactivity: A review and synthesis. *Psychological Bulletin, 95*, 387-409.

Sprague, R., & Sleator, E. (1977). Methylphenidate in hyperkinetic children: Differences in dose effects on learning and social behavior. *Science, 198*, 1274-1276.

Swanson, J., & Kinsbourne, M. (1979). The cognitive effects of stimulant drugs on hyperactive (inattentive) children. In G. Hale & M. Lewis (Eds.), *Attention and the development of cognitive skills*. New York: Plenum.

Varley, C. K. (1983). Effects of methylphenidate in adolescents with attention deficit disorder. *Journal of the American Academy of Child Psychiatry, 22*, 351-354.

Wells, K. C., Conners, C. K., Imber, L., & Delameter, A. (1981). Use of single-subject methodology in clinical decision-making with a hyperactive child on the psychiatric inpatient unit. *Behavioral Assessment, 3*, 359-370.

Wender, P. H., Reimherr, F. W., Wood, D., & Ward, M. (1985). Controlled study of methylphenidate in treatment of attention deficit disorder, residual type, in adults. *The American Journal of Psychiatry, 142*, 547-552.

What Do We Know About the Use and Effects of Behavior Therapies in the Treatment of ADD?

Karen C. Wells, PhD

ABSTRACT. Pharmacological intervention using stimulant medication has been shown to be an effective, but not maximally effective, treatment for Attention Deficit Disorder. In addition, several other limitations of stimulants when used as the sole treatment have resulted in the continued investigation of alternative and/or adjunct therapies for ADD children. The most viable of these is behavior therapy. Several techniques involved in the behavioral treatment of ADD are reviewed, including parent training, classroom management strategies, and social skills training. Like stimulant medication, behavior therapy has been found to be effective, but not maximally effective, in treating the full range of symptoms in the syndrome. The effects of behavior therapy include improvement in both primary (inattention, distractibility, impulsivity, hyperactivity) and secondary (aggression, conduct problems) symptoms. Only when behavior therapy and stimulants are combined, however, is full normalization achieved. A case of an ADD child is presented that illustrates the combined therapy approach and the single-subject experimental methodology in evaluating clinical outcome.

Pharmacological intervention using stimulant medications is the most common and accepted treatment of Attention Deficit Disorder. It is also the class of intervention on which the most research evaluation has been done, and there is now little doubt regarding the short-term efficacy of stimulant medication for children with ADD (Gittelman, 1983).

Karen C. Wells is Associate Professor of Psychiatry, Behavioral Sciences and Child Health and Development at George Washington University Medical School and is Director of Pediatric Psychology at Children's Hospital National Medical Center, 111 Michigan Avenue, N.W., Washington, D.C. 20010. Reprint requests should be addressed to the author at this address.

© 1987 by The Haworth Press, Inc. All rights reserved.

In spite of this widespread acceptance, there are also certain limitations of psychostimulant treatment when it is employed as the sole treatment for ADD (Pelham & Murphy, 1986). First, although most children improve with treatment, many do not achieve maximal improvement with medication alone. Several studies have shown that during treatment, symptoms remain significantly above the normative mean on both rating scales and behavioral observations (Gittelman et al., 1980). Second, stimulant medication often does not maximally affect the full range of symptomatology of hyperkinetic children. While positive effects are often found on measures of attention and activity and, to a lesser extent, on conduct problems (Solanto & Conners, 1982) these effects are often not achieved on other characteristics of the syndrome such as academic underachievement and poor peer relationships (Conners & Wells, 1986).

A third limitation of stimulant medication has to do with its applicability to home behavior problems (Pelham & Murphy, 1986). Many clinicians discontinue the use of stimulant medications after school hours, on weekends and during the summer (non school) months in order to avoid growth and appetite suppression and other undesirable side effects. When no other form of treatment is provided, parents are left to their own devices to control disruptive, hyperactive behavior at home. They frequently resort to coercive, hostile and overly punitive interactions which may exacerbate rather than reduce the ADD child's behavior problems. This is a particularly unfortunate state of affairs given that child and parent aggression are among the greatest predictors of poor outcome (Milich & Loney, 1979).

Finally, psychostimulant medication is beneficial for only 60-70% of ADD children; fully 30-40% do not achieve a significant treatment effect in group outcome studies (Conners & Werry, 1979; Eisenberg & Conners, 1971). Furthermore, for those children who do achieve a significant short-term treatment effect, there is no evidence that stimulant medications improves the long term prognosis of ADD children (Riddle & Rapoport, 1976; Weiss, Kruger, Danielson & Elman, 1975).

For all of these reasons, other therapy modalities will continue to be employed with ADD children, notably behavioral approaches. Behavior therapy is second only to stimulant medication in terms of the number of studies appearing in the clinical research literature.

Based on this research, it is clear that behavior therapy approaches represent viable treatment alternatives or adjuncts to stimulant medications.

CLINICAL BEHAVIOR THERAPY WITH ADD

Mash and Dalby (1979) were among the first to note that "behavior therapy," as it has been used with ADD children, encompasses many technologies, including parent training, operant reinforcement of academic responses, relaxation training, token economies and behavior contracts, to name a few. The clinical treatment of these children usually involves a comprehensive treatment plan including intervention in the home, school and, sometimes, in the peer group as well. Let us review the elements of a comprehensive behavioral treatment plan followed by what we know about the effects of these interventions.

Parent Training

Training parents in behavior management strategies is an essential component of treatment for ADD children, whether or not they are also receiving stimulant medications. Parents have the primary responsibility for socializing their children, teaching them to curb impulses and to exercise self-control, all developmental tasks which are rendered more difficult by virtue of the ADD child's inherent problems with short attention span, impulsivity and distractibility. As noted earlier, parents often must deal with the child's behavior during weekends, evenings, holidays and summers when medication may not be administered. In addition, there is a high degree of reciprocity between the behavior of ADD children and their parents such that children's behavior elicits a coercive, controlling, critical parental style from the parents (Barkley & Cunningham, 1979). This coercive style can then feedback to the child and promote the escalation and maintenance of aggressive acting-out behavior in the child (Conners & Wells, 1986; Wells & Forehand, 1985). Parents must learn how to interrupt this coercive cycle of interaction in order to manage momentary disruptive behavior as well as to avoid long-term damage to their relationship with their child.

A parent training program that has been employed frequently

with parents of ADD children was developed by Constance Hanf, researched by Rex Forehand and his associates (Forehand & McMahon, 1981) and promoted by Barkley (1981). In this program, parents are taught, via discussion, modeling, role-playing and homework assignments, a set of parenting skills that can be used to increase the child's appropriate behavior and decrease inappropriate or excessive behaviors, notably noncompliance. The skills are introduced in hierarchical fashion such that progression to the next skill is contingent upon successful completion of the one before.

Parents are first taught how to pay attention to their child's good behavior using "attending and following" skills in which they describe the child's ongoing behavior out loud. Attending practice is done as a method for reinforcing appropriate child behavior, increasing the parents' reinforcement value to the child, and promoting a more positive relationship between parent and child.

After practice in attending, parents then learn how to give verbal and nonverbal rewards to their children. Rewards are then added to attending skills, and the parents practice attending and rewarding while the child plays at home 15 minutes a day. Once play sessions are going smoothly, parents learn to use their attending and rewarding skills to encourage appropriate child behavior (e.g., compliance to instructions, finishing a task) and to withdraw attention as a means of decreasing inappropriate behavior.

Following this first phase of therapy, parents learn a specific punishment technique, time-out, for noncompliance and other behaviors that cannot be ignored. First, however, parents learn to give clear, specific, direct, concise commands that are age appropriate and do not exceed the developmental capabilities of the child. They, then, are taught to implement a time-out technique contingent upon noncompliance to parent instructions. Time-out involves sitting the child in a chair in a doll corner in the house and ignoring all talking or misbehavior the child displays in the chair. After time-out, children are taken back to the original situation and given the initial instruction again. The time-out sequence continues until the child eventually complies to the command.

While the above parenting techniques seem relatively simple and straightforward, they are usually fairly complex and require a great deal of in-session practice and role playing for parents to implement them successfully at home. In addition, a great deal of clinical skill is necessary on the part of the therapist in working with parents. Parents first must learn about the nature of the ADD syndrome and the target behaviors that result from it. Presenting ADD as a developmental disorder often helps diffuse blame-oriented attributions that the parents have been making about the child and themselves. Similarly, "parent training" as a therapeutic modality carries with it a danger that parents will conclude that their poor parenting *caused* their child's disorder.

I frequently point out that raising an ADD child requires more than the ordinary level of parenting skills that most adults are intuitively equipped with, and that it is not uncommon in families for parents to raise successfully two or three normal siblings while having considerable difficulty parenting their ADD child. Such discussions frequently elicit emotional reactions from parents who have saddled themselves and each other with blame, guilt and anxiety and whose own mental health and marital relationships may have been affected. The clinician must be prepared to deal with these issues as they arise during the course of parent training.

Once the basic set of techniques is mastered, the therapist must then help the parents implement behavior change programs to address the idiosyncratic difficulties of their particular child. For example, in order to help the child develop rule governed behavior, the parents might develop a set of "house rules," the violation of which results in immediate time-outs. Likewise, the parents might develop shaping programs to help the child practice progressively longer periods of on-task behavior or accomplish more age appropriate expectations (e.g., cleaning up his room). Likewise, parents must be assisted in developing procedures for controlling the child's behavior outside the home. All of these strategies will be easier for parents to implement once they have learned to deal with the child's noncompliant behavior.

Behavior Therapy in the Classroom

Another major aspect of a multicomponent behavioral treatment plan is the use of behavioral strategies in the classroom. Indeed, there are many more research evaluations of the effects of behavioral intervention in the school setting than in the home.

1. Teacher consultation. Treatment often begins with a series of consultation sessions with the classroom teacher, the purposes of which are to instruct the teacher in principles and techniques of social learning theory and to enlist her assistance in devising the individualized plan. Some investigators (e.g., Pelham, 1982) have teachers read manuals on social learning theory and discussion revolves around how the teacher can praise and pay attention to appropriate behavior, ignore minor disruptive behavior, and use mild punishment (brief isolation) for disruption which cannot be ignored.

2. *Point systems.* The teacher and consultant develop a list of target behaviors that are relevant for the particular child, and that can be incorporated into school and/or home based token reinforcement systems. Typical target behaviors have included accurate completion of academic work and staying on-task as well as behaviors to be decreased (e.g., out-of-seat without permission, noncompliance, calling out in class, not listening). Pelham (1982) has indicated that with hyperactive children it is often necessary to have points exchangeable for consequences delivered at school (e.g., free time as soon as work is completed) as well as at home. In this regard, the therapist can negotiate a Daily Report Card system in which the teacher checks off on an index card the child's performance on target behaviors. The child takes this card home where backup consequences are delivered by parents.

3. *Response cost.* Recent work has indicated the potential importance of including response cost procedures in classroom behavioral programs for ADD children. In one study that has evaluated this procedure, response cost added significantly to the treatment effect obtained with a standard behavioral program (Atkins, White, Adams, Case & Pelham, 1984). In this study, each occurrence of a targeted inappropriate behavior resulted in the loss of one minute of play time for that day. In addition, rewards were given for days in which the child lost three or fewer points. In another study (Schell et al., 1983), each child began the class period with 100 points that were exchangeable after class for toys. During each class period, the teachers set a timer for varying intervals. When the timer sounded, children who were on task kept their 100 points. Those who were off task lost a proportion of their points for each occurrence. One could conceive of a system combining positive reinforcement and response cost in which points are earned for performance of positive target behaviors (bringing in homework, completing academic assignments, on-task) and points are lost for inappropriate behaviors (noncompliance, out-of-seat etc.).

Social Skills Training

A relatively new innovation in behavior therapy for ADD is social skills training. The need for intervention in the social domain arises from observations that the peer relations of ADD children are very seriously disturbed, that early peer problems predict later maladjustment to a very significant extent (Conners & Wells, 1986) and that standard behavioral interventions do not significantly improve ADD children's social adjustment, (e.g., Pelham, Schnedler, Bologna & Conteras, 1980).

Several approaches to social skills training are currently being investigated (Bogart & Wells, 1985; Pelham et al., 1979; Hinshaw, Henker & Whalen, 1984). Perhaps the most naturalistic approach involves training in the context of peer play groups. In this approach, the skills and concepts of communication, participation, and cooperation are first addressed in small group sessions using instructions, modeling, and role-playing techniques. Instructional sessions are followed by play sessions in which children practice skills newly acquired while playing age-appropriate group games. They earn points for appropriate play and for practicing skills learned in instructional sessions. Points are exchangeable for tangible rewards. Because the poor peer relations of ADD children are so resistant to change with standard or short-term treatments, it will undoubtedly be necessary to involve ADD children in structured social skills groups over a long period of time in order to have an appreciable impact on this domain.

THE EFFECTS OF BEHAVIOR THERAPY WITH ADD

It is impossible to make a sweeping generalization about the effects of behavior therapy with ADD because studies evaluating efficacy have varied greatly with respect to the subject characteristics of ADD children, the adequacy and comprehensiveness of the behavioral treatment employed, whether or not stimulant medications were also used, and the nature and comprehensiveness of the outcome measures. In addition, treatment outcome studies have usually found a high degree of variability in response of subjects in the same study to the same treatment, probably reflecting the heterogeneity of what we globally refer to as Attention Deficit Disorder. Stated simply, this means that different ADD children will respond to the same treatment(s) differently.

With these provisos in mind, we can conservatively state that behavior therapy, when properly administered, is more or less effective for a large number of ADD children on a number of symptom measures, and it is the most viable alternative and/or adjunct to stimulant medications at the present time (Conners & Wells, 1986).

Studies have shown significant effects of behavior therapy in the classroom on ADD children's off-task behavior, inattention, distractibility, academic performance, hyperactive motor behavior and on teacher ratings of conduct problems in the classroom such as aggression and disrupting others (e.g., Gittelman et al., 1980; O'Leary, Pelham, Rosenbaum & Price, 1976; Pelham et al., 1980; Wells, Conners, Imber & Delamater, 1981). Studies that also include intervention in the home (e.g., parent training) and measures of home behavior have found significant effects on such measures as daily problem behavior ratings by parents, parent ratings of activity level and aggression in the home, parent global ratings of improvement as well as specific observed behaviors such as child compliance to parental commands and parents use of positive social attention following child compliance (e.g., Gittelman et al., 1980; Pelham et al., 1980; Pollard, Ward & Barkley, 1984).

The effects of *standard* behavior therapy (i.e., in the home and classroom) on the peer relationships and social adjustment of ADD children have been less compelling. For example, in a series of studies Pelham has found that *after* treatment, ADD children still averaged two standard deviations above their class means on sociometric measures of peer acceptance (Pelham & Bender, 1982). This has prompted a few recent studies in which social skills training has been added to standard behavioral treatment. Even with such treatment, results continue to be somewhat disappointing. For example, Pelham et al. (1979) found no *statistically* significant effect on negative peer nominations of adding social skills training to standard behavior therapy. Nevertheless, those children who received the training showed decreases in negative nominations twice as great as those who did not receive the training. These findings, together with the fact that stimulant medication does not significantly improve the peer relations of ADD children, provide some indication that social skills training should continue to be investigated.

While behavior therapy has been shown to be effective in treating ADD as summarized above, recent controlled studies have shown that, for most children, it is not maximally effective. This conclusion is suggested by studies showing that after behavioral treatment in which good clinical improvements have been obtained, the ADD children are nevertheless still significantly more deviant than *normal* classmate controls on a number of crucial symptoms (Gittelman et al., 1980; Pelham et al., 1980). In addition, a number of studies have shown an incremental effect of adding stimulant medication to behavior therapy such that the effects obtained with the combined approach are greater and more normalizing than the effects obtained with either treatment alone (Gittelman et al., 1980; Pelham et al., 1980; Wells et al., 1981). Thus, neither stimulants alone nor behavior therapy alone represents maximal treatment for ADD children. When they are combined, however, many children will achieve normal levels of functioning on many symptom measures.

Because of the unsatisfactory results from the use of behavior therapy or stimulants alone, and because of the highly individualized response of ADD children to various treatments, it is necessary for clinicians and researchers to have a method for establishing the most effective treatment(s) for any given child. Single subject experimental designs represent a powerful methodology for determining the effects of treatment in an individual, combining the best features of individualized clinical management with experimental control and objective assessment (Conners & Wells, 1982; Wells et al., 1981). The following case example illustrates the use of single subject methodology in determining the best treatment for a particular child.

Randy was a 9 year-old child with severe ADD and hyperactivity referred for inpatient evaluation and treatment by his public school special education teacher and principal. Randy displayed very short attention span and concentration in class, excessive motor behavior, and aggression towards peers and adults.

Following collection of baseline measures by a trained, objective observer in the unit classroom, Randy was started on 5 mg., twice per day of Dexadrine with no appreciable improvement (See Phase 2, Figure 1). Following a brief return to baseline (Phase 3), methylphenidate was begun, titrated to 15 mg. twice per day (Phase 4) with clear improvement in off-task and gross motor behavior. The addition of a behavioral program in the classroom, in which Randy learned to give himself tokens whenever a tone sounded and he was on task, resulted in further improvement in off-task and gross motor behavior, greater than that achieved with Ritalin alone. The withdrawal of medication (Phase 6) followed by its reinstatement in Phase 7 verified that the combined approach was more effective than either medication alone (Phase 4) or behavioral treatment with placebo (Phase 6). Randy was discharged from the hospital with a recommendation that methylphenidate and behavioral programs continue to be used in his special school classroom.

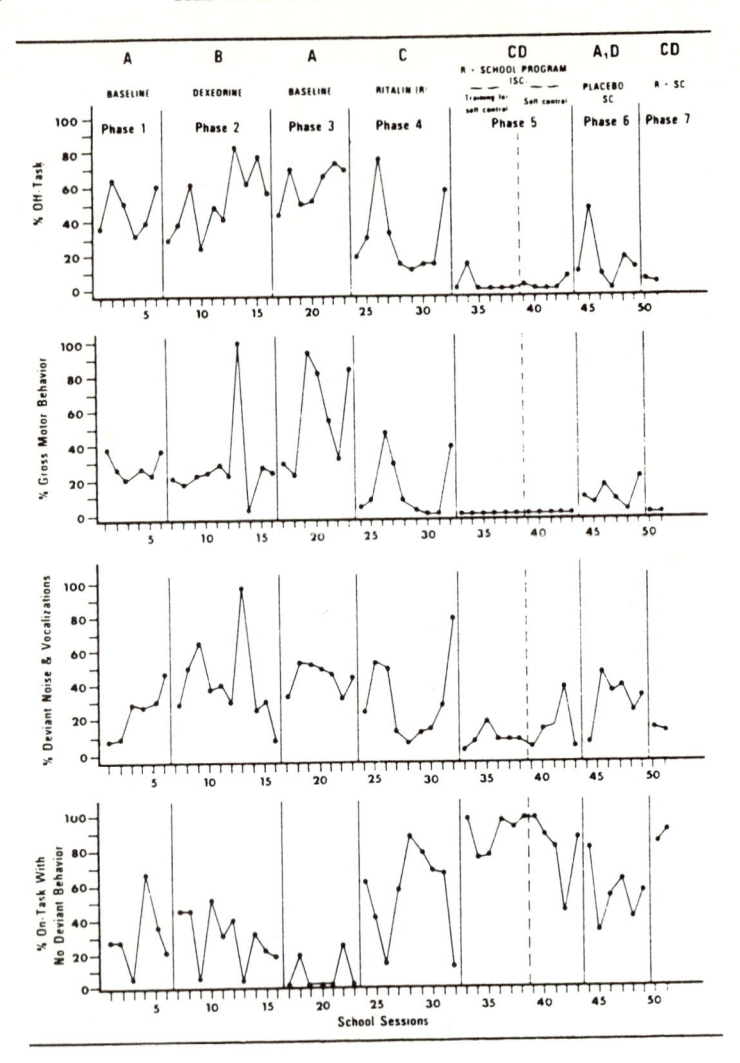

FIGURE 1. Percent occurrence in the classroom of off-task behavior, gross motor behavior, deviant noise and vocalizations, and on-task with no deviant behavior, measured across baseline, medication, school behavioral program, and combination treatment phases.
Note. From "Use of single subject methodology in clinical decision making with a hyperactive child on the psychiatric inpatient unit" by K. C. Wells, C. K. Conners, L. Imber, and A. Delamater, 1981, *Behavioral Assessment, 50*, p. 365. Copyright 1981 by the Association for Advancement of Behavior Therapy. Reprinted by permission.

REFERENCES

Atkins, M.S., White, K.J., Adams, P.N., Case, D.E., & Pelham, W.E. (1984). Behavioral and pharmacological treatment of a hyperactive aggressive child. Unpublished manuscript, Florida State University.

Bogart, C., & Wells, K.C. (1985). *Social skills training for children with Attention Deficit Disorder and Hyperactivity*. Unpublished manuscript, Children's Hospital National Medical Center, Washington, DC.

Barkley, R.A. (1981). *Hyperactive children: A handbook for diagnosis and treatment*. New York: Guilford Press.

Barkley, R.A., & Cunningham, C.E. (1979). The effects of methylphenidate on the mother-child interactions of hyperactive children. *Archives of General Psychiatry, 36*, 201-208.

Conners, C.K., & Wells, K.C. (1982). Single case designs in psychopharmacology. In A.E. Kazdin & A.H. Tuma (Eds.), *New directions for methodology of social and behavioral science: Single-case research designs*. San Francisco: Jossey-Bass.

Conners, C.K., & Wells, K.C. (1986). *Hyperkinetic children: A neuropsychosocial approach*. Beverly Hills: Sage.

Conners, C.K., & Werry, J.S. (1979). Pharmacotherapy. In H.C. Quay & J.S. Werry (Eds.), *Psychopathological disorders of childhood* (2nd ed.). New York: Wiley.

Eisenberg, L., & Conners, C.K. (1971). Psychopharmacology in childhood. In N. Talbot, J. Kogan, & L. Eisenberg (Eds.), *Behavioral science in pediatric medicine*. Philadelphia: Saunders.

Forehand, R., & McMahon, R.J. (1981). *Helping the noncompliant child: A clinician's guide to parent training*. New York: Guilford.

Gittelman, R. (1983). Hyperkinetic syndrome: Treatment issues and principles. In M. Rutter (Ed.), *Developmental neuropsychiatry*. New York: Guilford.

Gittelman, R., Abikoff, H., Pollack, E., Klein, D.F., Katz, S., & Mattes, J. (1980). A controlled trial of behavior modification and methylphenidate in hyperactive children. In C.K. Whalen & B. Henker (Eds.), *Hyperactive children: The social ecology of identification and treatment*. New York: Academic Press.

Hinshaw, S.P., Henker, B., & Whalen, C.K. (1984). Self-control in hyperactive boys in anger-inducing situations: Effects of cognitive-behavioral training and of methylphenidate *Journal of Abnormal Child Psychology, 12*, 55-77.

Mash, E.J., & Dalby, J.T. (1979). Behavioral interventions for hyperactivity. In R.L. Trites (Ed.), *Hyperactivity in children: Etiology, measurement, and treatment implications*. Baltimore: University Park Press.

Milich, R.S., & Loney, J. (1979). The role of hyperactive and aggressive symptomatology in predicting adolescent outcome among hyperactive children. *Journal of Pediatric Psychology, 4*, 93-112.

O'Leary, K.D., Pelham, W.E., Rosenbaum, A., & Price, G.H. (1976). Behavioral treatment of hyperkinetic children: An experimental evaluation of its usefulness. *Clinical Pediatrics, 15*, 510-515.

Pelham, W.E. (1982). Childhood hyperactivity: Diagnosis, etiology, nature and treatment. In R. Gatchel, A. Baum, & J. Singer (Eds.), *Behavioral medicine and clinical psychology: Overlapping disciplines*. Hillsdale NJ: Erlbaum.

Pelham, W.E., & Bender, J. (1982). Peer relations in hyperactive children: Description and treatment. In K.D. Gadow & I. Bialer (Eds.), *Advances in learning and behavioral disabilities* (Vol. 1). Greenwich, CT: JAI Press.

Pelham, W.E., & Murphy, H.A. (1986). Attention deficit and conduct disorders. In M. Hersen (Ed.), *Pharmacological and behavioral treatments: An integrative approach*. New York: Wiley.

Pelham, W.E., Schnedler, R.W., Bologna, N.C., & Conteras, J.A. (1980). Behavioral and stimulant treatment of hyperactive children. A therapy study with methylphenidate probes in a within-subject design. *Journal of Applied Behavior Analysis, 13*, 221-236.

Pelham, W.E., Schnedler, R.W., Miller, J., Ronnei, M., Paluchowski, C., Budrow, M., Marks, D., Nilsson, D., & Bender, M. (1979). *The combination of behavior therapy and psychostimulant medication in the treatment of hyperactive children: A therapy outcome study.* Paper presented at the meeting of the Association for Advancement of Behavior Therapy, San Francisco, CA.

Pollard, S., Ward, E.M. & Barkley, R.A. (1984). The effects of parent training and Ritalin on the parent-child interactions of hyperactive boys. *Child & Family Behavior Therapy, 5,* 51-69.

Riddle, K.D., & Rapoport, J.L. (1976). A two-year follow-up of 72 hyperactive boys. *Journal of Nervous and Mental Disease, 162,* 126-134.

Schell, R., Pelham, W.E., Adams, P., Atkins, M., Greenstein, J., White, K., Bender, M., Bailey, J., Shapiro, S., Law, T., Darling, J., & Case, D. (1983). *The effects of a response-cost program on classroom behavior of hyperactive children.* Paper presented at the annual meeting of the Association for Behavior Analysis, Milwaukee, WI.

Solanto, M.V., & Conners, C.K. (1982). A dose-response and time-action analysis of autonomic and behavioral effects of methylphenidate in attention deficit disorder with hyperactivity. *Psychophysiology, 19,* 657-658.

Weiss, G., Kruger, E., Danielson, U., & Elman, M. (1975). Effects of long-term treatment of hyperactive children with methylphenidate. *Canadian Medical Association Journal, 112,* 159-165.

Wells, K.C., Conners, C.K., Imber, L., & Delamater, A. (1981). Use of single-subject methodology in clinical decision-making with a hyperactive child on the psychiatric inpatient unit. *Behavioral Assessment, 3,* 359-369.

Wells, K.C., & Forehand, R. (1985). Conduct and oppositional disorders. In P.H. Bornstein & A.E. Kazdin (Eds.), *Handbook of clinical behavior therapy with children.* New York: Dorsey Press.

Cognitive Behavior Therapy for Hyperactive Children: What Do We Know?

Carol K. Whalen, PhD
Barbara Henker, PhD

ABSTRACT. The background, rationale, and status of cognitive-behavior therapies for hyperactive children are summarized. Emerging themes and redirections are identified, including attributional foci, combination therapies, and tailored treatments. Dilemmas for the scientist-practitioner are highlighted, as the uneven track record of tangible improvements is balanced against the potential for less tangible (and more difficult to measure) gains in causal reasoning, self-perceived efficacy, and response to challenge. The article concludes with a set of guidelines and recommendations for cognitive-behavior therapies with hyperactive children.

BACKGROUND AND BASIC TENETS

Cognitive-behavior therapy (CBT) is a generic term applied to a protean brew of strategies and techniques. Underlying this heterogeneity of therapeutic targets and procedures, however, are a few interrelated premises. One is the centrality of cognitive representations (thoughts, beliefs, and attitudes) based, in turn, on the notion that cognitions are behaviors in their own right. A second is the assumption of bidirectional causality: Cognitions actively mediate—and are also modified by—both affect and overt action. In other words, thoughts are actions that represent integrations of past and ongoing experiences and also serve as self-propellers for subsequent events. A third proposition is that self-perceptions and self-guiding strategies, like behaviors, can and should be direct targets for change. And finally, CBT approaches presume that a functional

---

Carol K. Whalen, Social Ecology, University of California, Irvine, CA 92717. Barbara Henker, Department of Psychology, University of California, Los Angeles, CA 90024.

© 1987 by The Haworth Press, Inc. All rights reserved.

understanding of psychopathology and treatment depends on an understanding of cognitive processes and products (Kendall, 1985; Mahoney, 1977). Given these core assumptions, CBT serves as a balancing force in the world of child therapy, offsetting both the "black box" environmentalism that characterizes many operant approaches and the mechanistic and reductionistic emphases that often pervade pharmacologic treatments. CBT recognizes the important contributions of cognition to behavior problems and to their solutions. But unlike the traditional psychotherapies, which also focus on cognition, CBT takes a scientific approach, insisting that cognitions or covert behaviors, like overt behaviors, are subject to systematic analysis and evaluation. In fact, intervention and evaluation are considered inextricably enmeshed, and no CBT program can be complete without both components.

There is a natural affinity between CBT and ADD[1] (Whalen, Henker & Hinshaw, 1985). Hyperactive children have pervasive, high-impact difficulties, many of which seem to persist or intensify rather than to diminish over time. By middle childhood, most hyperactive youngsters and their families have undergone diverse attempts at amelioration. Stimulant medication and behavior modification may be quite effective in the short run, but often with limited generality and durability. Regardless of their treatment histories, the majority of hyperactive children continue to be described as impulsive, noncompliant, and unable to inhibit or delay. Some of the primary components of CBT, including careful planning and reflective problem solving, would appear to be perfect antidotes. A two-pronged focus on self-monitoring and self-management seems ideally suited to a child who shows behavioral excesses and has difficulty modulating his actions in accordance with changing environmental cues and demands. Thus the pervasiveness and recalcitrance of hyperactive children's problems, the limitations of existing interventions, and the conceptual coherence between the deficits and the treatments all ensured that CBT would be adopted readily by those struggling to help hyperactive children.

Since the initial demonstrations of CBT in the early 1970s (e.g., Meichenbaum & Goodman, 1971), numerous ADD treatment programs have been launched with unbridled energy and enthusiasm. The primary focus has been on teaching these children strategies for self-regulation—to talk to themselves in order to guide their behaviors. Children are taught first to "think aloud" so that their internal monologues can be monitored and modified by the therapist. As they progress, they are encouraged to convert overt talking into covert thinking. This prevents their "cognitive prosthetics" from disturbing others, thus making them ready targets for teasing. CBT usually involves several additional ingredients to maximize both the occurrence and the impact of self-instruction. These include modeling, problem-attack strategies, techniques for detecting errors and generating alternative solutions, overt and covert rehearsal, goal setting, self-appraisal and self-reinforcement, positive reinforcement from the trainer, and perhaps even response cost. Thus the boundaries of CBT as practiced are somewhat blurred; most programs present multipurpose packages that contain conventional behavioral and tutorial ingredients as well as specific cognitive-behavioral components. In many ways, CBT consists of standard educational and behavioral techniques with a cognitive overlay. (Additional details about CBT programs for children can be found in Kendall and Braswell, 1984, and in Meyers and Craighead, 1984.)

Although notable successes have been documented, the experience gained with CBT for ADD during the past 15 years has led to disappointment and even disenchantment, as well as to significant changes in emphasis and direction. We begin with a characterization of early misconceptions and emerging trends, followed by an analysis of CBT outcomes and a specification of new challenges for the scientist-practitioner. The final section provides guidelines and specific recommendations for optimizing CBT approaches to hyperactivity.

RETIRING ASSUMPTIONS AND EMERGING THEMES

CBT for ADD Is More Than Self-Instructional Training

As described above, the teaching and correcting of internal monologues—the self-talk people use to plan, propel, and review their behavior—is a core component of CBT. The impulsivity and repeated failures of hyperactive children have led to the assumption that their self-talk skills are deficient, which often appears to be the case. What tends to be forgotten, however, is that self-instructional training is only one of several CBT approaches that may be indicated for optimal therapeutic impact. Children with impulse control problems may indeed benefit from planful, systematic self-talk designed to slow behavioral enactment processes to allow time to think, plan, and evaluate. But significant gains may also accrue following different or additional cognitive approaches. One intriguing example is attribution or reattribution training, an approach that has not yet been applied or tested extensively in clinical treatment programs for children.

The Role of Causal Attributions

Consider a situation in which two children have just failed a test at school. One believes that his poor grade resulted from a lack of effort, while the other thinks that his teacher makes impossible tests, and that he's not really very smart anyway. Who will be likely to study harder before the next exam? To take an example from the social realm, a child's level of prosocial activity following a rebuke from a peer may depend, at least partially, on whether he attributes the rebuke to his own lack of friendliness or to hostility on the part of the other child. Analogously, a child who believes that an ambiguous event (such as getting hit in the back with a ball on the playground) was a willful act on another child's part, rather than an accident, may be more likely than his peers to show high levels of aggression (Dodge & Frame, 1982).

We know that children, like adults, are neither passive nor mindless targets of life events. Rather, they interpret and frame their experiences, actively seeking explanations and building theories of human behavior. The results of these attempts to answer the question "why" may influence a child's subsequent behavior patterns, as well as his affect, motivation, perceptions of self-efficacy, and expectancies about future events and outcomes. When a child's causal reasoning appears linked to maladaptive or avoidant behavior, direct attempts to change attributional processes may prove beneficial.

Attribution and Reattribution Training

There are several ways in which attribution training may facilitate long-range objectives for hyperactive children. At this point, it seems useful to distinguish between antecedent and program-generated attributional processes (Henker, Whalen & Hinshaw, 1980).

Antecedent attributions are constructs that the child brings to the clinical setting. Included here are causal explanations of his own behaviors and of the actions of other people. Whether or not these attributional accounts are veridical, they may be considered change-worthy when they impede optimal functioning. In terms of the child's own behavior, the most common focus is on effort attributions, the assumption being that children will be more likely to attempt to change and to succeed if they ascribe past problems to an internal and modifiable cause—insufficient effort on their own part—than if they consider the basic cause to be something that is either enduring or beyond their own control (e.g., a biological deficit or the actions of another person). In terms of the child's analyses of other peoples' behaviors, a goal may be to encourage him to reattribute ambiguous events to benign or chance factors, rather than to assume that they resulted from willful negativity on the part of another person.

Program-generated attributions involve the surplus meanings attached to various diagnostic practices and treatments. It is often assumed, for example, that children who are diagnosed ADD and especially those who are prescribed psychoactive medication have a biological deficit. This assumption may encourage some parents and teachers to decrease their expectations and may relieve the child of responsibility for his own actions, a situation that can be accepted readily by a perceptive child who wish to avoid work or blame.

Rather than leave these inadvertent messages to chance, therapists can introduce treatment programs in ways that counteract potentially undesirable attributions. Labeling psychoactive medication as a "magic pill" or an "arithmetic pill" would be expected to have quite a different impact than describing it as "a crutch that will help you but will not do the walking for you." Therapists can also attend to the cognitive processes of the child's significant others, encouraging the view that the intervention is a temporary aid for the child rather than confirmatory evidence of deficiency. It may also be helpful to initiate self-control training prior to or in conjunction with a more externally-focused behavioral or medication program, so that children will have an internal attributional anchor—their own developing self-regulatory competencies—for the positive behavior changes that result. Without such internal anchors, a child may worry that his performance will deteriorate as soon as the pills or the tokens are withdrawn.

Preventing or correcting maladaptive attributional accounts cannot be expected to ameliorate well-entrenched problems or teach needed skills. There are, however, several potential beneficial outcomes. Enhancing a child's perceptions of causal linkages between his own behaviors and outcomes may (a) protect self-esteem and heighten self-perceived efficacy; (b) increase persistence and perseverance; (c) increase expectancies for success and the probability of accepting difficult tasks and new challenges; and (d) prevent over-reliance on a therapist or therapeutic agent and thereby facilitate treatment discontinuation.

An illustrative example of the use of reattribution training to facilitate medication withdrawal is presented by Rosén, O'Leary, and Conway (1985). A 9-year-old boy who was functioning well in a behavioral program showed marked deterioration when stimulant medication was discontinued. He made comments such as, "My pills make me get done with the work" and "I get angry without my pill." Given the resurgence of behavior problems, the medication regimen was reinstated, but this time, unbeknownst to the child, the pills were placebos. At the same time, the teacher was recruited as a "reattribution therapist," emphasizing repeatedly to the child that he, and not the medication, controlled his behavior. This combined approach resulted in marked behavioral improvement. The child was then allowed to earn his way off (placebo) medication, after which he continued to perform well with neither medication nor placebo.

It is important to recognize, however, that these potential outcomes assume matching levels of skill development and success. If a child consistently fails in academic or social activities, reattribution training of the type described above can do more harm than good, by reinforcing notions of deficiency and encouraging self-blame. In other words, competency must be ensured before attributions of personal responsibility can be expected to have a therapeutic impact.

Not much is known about the relative potency of attribution interventions, or even about linkages among attributions, affect, and behaviors. Interrelationships are still a mystery even within the cognitive realm; for example, little is known about the links among causal ascriptions, self-perceived efficacy, and expectancies for success. Nor do we know very much about some logically prior questions, including how attributions develop and how malleable they are. Moreover, the range of attributional interventions is quite restricted at present, and many of the programs were designed for adults rather than for children (Försterling, 1985). But despite these limitations in our knowledge and experience, attributional approaches to intervention merit exploration and testing. Both the experimental and the clinical literature indicate their potential value in the therapeutic armamentarium for hyperactive children.

Assumptions of Exclusivity and the Trend Toward Combination Therapies

Many clinicians have assumed that, because of its focus on "higher" (metacognitive) processes, as well as its conceptual match with the putative deficits of hyperactive children, CBT can and should replace more traditional behavioral, educational, and pharmacologic routes to remediation. There is a soft and a hard version of this "exclusivity myth." The soft version is based partially on the early enthusiasm surrounding CBT, which was spurred by growing disenchantment with existing approaches. CBT focuses on the "fundamentals" for daily living, and thus additional amelioration seems unnecessary. The child is expected to learn "portable" coping strategies that can be transferred from setting to setting, thus eliminating dependence on external monitoring and intervention and facilitating that most elusive of treatment objectives: generalization. The hard version grapples with a more basic issue: the belief that the core tenets taught by CBT may be inherently incompatible with those implicit in behavioral and pharmacologic approaches. CBT proposes to produce self-direction and self-control, while medication and behavior modification rely on external agents of behavior change.

Thus, one thought was that CBT obviates the need for other approaches, while another was that the inclusion of these other approaches might actually impede progress. This exclusivity myth is a curious phenomenon because, as noted above, the earliest applications of CBT with hyperactive children clearly included standard components of behavioral treatment such as modeling and reinforcement. In many instances, the most appropriate intervention might consist of old-fashioned demonstration and practice, with hearty doses of corrective feedback and reinforcement, along with self-instructional training. There are also indications that response cost may facilitate CBT. Fortunately, both versions of this exclusivity myth are beginning to fade with the recognition that CBT is not a panacea in either precept or practice, and that there are far better justifications for combined than for competing treatments (Douglas, 1983; Hinshaw, Henker & Whalen, 1984a; Meichenbaum, 1985).

A Multipurpose Package versus Tailored Treatments: Specifying Specificity

For some cognitive behavior therapists, the goal is to develop an intervention program that systematically treats academic, social, and other difficulties of hyperactive children in a prescribed sequence of lesson plans or training sessions. This is considered feasible and cost-effective because the hyperactive child's presumed underlying deficits in self-regulatory skills pervade all performance domains. Some disappointing results in recent years (see below) have led to a reconsideration of these assumptions and growing recognition of the need for tailored treatments (e.g., Abikoff, 1985; Mahoney & Nezworski, 1985). There are at least three important dimensions to consider when moving toward specificity: individual differences, response domains or modalities, and cognitive-behavioral processes.

Individual Differences

Although it is well known that hyperactive children form a highly heterogeneous group, individual and developmental differences are rarely considered when intervention programs are designed. Only recently have cautions and caveats begun to appear in the literature, indicating that there is no single CBT that can be all things for all children. Numerous dimensions have been identified as potentially relevant to CBT efficacy, including developmental level, IQ, verbal skills, severity of disorder, response to reinforcement contingencies, preexisting modes of problem solving, spontaneous use of self-talk, tendencies toward autocriticism, attributional styles and locus of control, motivation to change, and family and environmental resources.

Before implementing a full-scale CBT program for an individual child, it is important to ensure not only that the child understands and can execute the procedures, but also that he is willing to do so. There are some preliminary indications that CBT may be best suited for children low on dimensions of problem severity and high on such dimensions as cognitive capacity, verbal skill, internal locus of control, autocriticism, and motivation to change. This is a familiar formula that has surfaced repeatedly in outcome studies of diverse therapeutic modalities. It merely suggests, once again, that the most competent and least disturbed individuals are likely to show the greatest treatment-related gains.

Response Domains and Modalities

Markedly similar cognitive-behavioral procedures have been proposed for a panoply of domains, in attempts to enhance selective attention, sustained attention, short-term memory, reflective decision making, spelling, reading, arithmetic, moral reasoning, rule-governed social behavior, social information processing, and interpersonal competence. These experiences have led clinical investigators to question the assumption that one set of problem-solving strategies can be acquired and applied across tasks and settings. It is becoming increasingly apparent that the search for context-free competencies may in fact be futile (Tharp & Gallimore, 1985). Not much is known yet about the best fit between procedure and domain, although there are preliminary indications that CBT may be particularly effective with cognitive impulsivity and deficient self-observation, while markedly less effective with noncompliance and social disruption.

Cognitive-Behavioral Processes

When we try to enumerate the component processes involved in effective problem solving, the list rapidly becomes unwieldy. To solve any problem, a child must recognize its existence and mobilize his resources, deploy his attention appropriately and sustain it over time, monitor his understanding and actions, generate alternative plans and solutions, arrange his environment to facilitate his efforts, maintain his own interest and motivation, appraise his performance accurately and modulate his behavior accordingly (cf. Whalen et al., 1985). Deficient performance can be due to problems in any link of this multidimensional chain or to some unknown combination of difficulties, and different hyperactive children will have different types and subsets of problems. Thus any general purpose CBT package is likely to miss the mark for an individual child, perhaps focusing too heavily on nonproblem areas and devoting insufficient attention to the child's most serious needs.

To summarize, there are multiple dimensions that may relate to the success of CBT with an individual child and for a specific problem. At this point, our assessment tools for measuring these dimensions remain rather primitive, as does our knowledge about how best to match treatment procedures with individuals, problem domains, and processes. Even without this information, however, the skillful clinician can attempt to assess child competencies and task requirements and link these dimensions to CBT outcomes. Such analyses, though imperfect, should markedly enhance the efficacy of CBT. In other words, the clinician can proceed on a case-by-case basis, helping to generate, as well as awaiting the needed research evidence.

Palatability, Safety, and Consumer Satisfaction: Needed Additions to the Decision Matrix

One reason, if not the major reason, for the enthusiastic reception given CBT is that this approach is assumed almost universally "palatable" and, indeed, ethically preferable, especially when contrasted to behavioral and pharmacologic interventions. The basic tenets of CBT, especially the focus on self-direction and internal control, are compatible with humanistic conceptions of individual freedom, developmental theories of the shifts from external to internal control during the school-age years, and current trends toward selecting the least restrictive alternatives in the care, education, and treatment of people who are different. In other words, CBT presents a basically wholesome treatment.

Fifteen years of experience with this approach, however, have indicated at least the possibility of unanticipated negative sequelae (Whalen et al., 1985). The many facets that need to be considered when making treatment determinations were discussed above, including the child's characteristics, type of task or activity, practical resources and restrictions, and psychosocial contexts. For example, CBT may be contraindicated for children who are highly anxious or perhaps excessively self-focused. For any child, careful cognitive planning and review may detract from performance on tasks that require speed or definitive action (Abikoff & Gittelman, 1985). More generally, the reflective techniques may be so cumbersome that the child's attention is diverted from the requirements of the task itself.

Another consideration is that CBT communicates to both the child and the caretakers the child's responsibility for her outcomes. This type of message can have salutary consequences, increasing a child's self-perceived efficacy and sense of responsibility. The child begins to take appropriate credit for her own accomplishments, and this often has an impact on the way others view her. As noted above, however, these positive effects presuppose that the child's competencies are developing apace with increasing self-expectations, and increased external demands regarding behavior. If expectations outstrip needed skill acquisition and behavior changes, the message of CBT can actually result in maladaptive frustration, blame, and guilt, rather than the desired increases in responsible behavior. Such inadvertent "emanative effects" or psychosocial sequelae may attend any type of treatment, including pharmacotherapies, behavior therapies, and psychotherapies (Henker & Whalen, 1980; Henker et al., 1980). The point here is not that such effects are specific to CBT, but rather that their impact may be obscured because of the assumption that CBT is, almost by definition, benign.

There are also reasons to question the assumption of universal palatability. CBT requires more time and effort from parents, teachers, and therapists than do other standard approaches, and even when the results are positive, behavior changes may not be as readily observed. It is also more difficult and at times more tedious for the child to repeatedly execute prescribed cognitive routines than to ingest pills or adjust to new behavioral contingencies. Moreover, there may be embarrassing aspects of CBT, such as the need for children to talk aloud during the early phases of some of the training modules, a component that has met with active resistance. The program must be palatable to parents if the treatment is to continue and to the child if he is to participate as actively and consistently as CBT requires. A less-than-enthusiastic child may still benefit from medication or systematic behavioral contingencies, but CBT is certain to fail if the child chooses not to engage in the prescribed cognitive strategies.

Consumer satisfaction has only recently become an issue in the realm of psychological and behavioral therapies. We do not yet know how or whether palatability and satisfaction are related to outcome for individuals who participate in particular therapeutic programs. What seems almost certain, however, is that these perceptions and preferences relate to willingness to continue a treatment program. The central point here is that negative effects must be monitored and palatability and consumer satisfaction evaluated rather than assumed. This prescription applies to all treatment modalities, but it may be most easily overlooked with CBT because of the humanistic premises that underly this approach and its clear compatibility with ethical precepts.

THE CONTRIBUTIONS OF CBT FOR ADD: THE DATA DILEMMA

The Many Faces of Efficacy and Accountability

Cognitive behavioral intervention programs for hyperactive or impulsive children have demonstrated some notable successes, but the gains documented to date tend to be short-term and circumscribed, or difficult to replicate. In some ways, the track record for CBT is similar to that for stimulant therapies and conventional behavioral approaches. Each has proven useful in a limited manner, but none has effected behavior changes of sufficient breadth, depth, and durability to enable the majority of these children to overcome or outgrow their diagnosis. All of these treatments fall short when applied to clinically diagnosed youngsters, particularly those with severe behavior disorders. Each approach is useful under certain conditions, with certain children, and for certain problem domains, but information about how to optimize the match between child and treatment is sorely lacking. Nor do we know how to balance potential gains against costs and risks.

Despite this similarity between CBT and alternative approaches, there seems to be a growing disenchantment with cognitive strategies, most likely due to the fact that expectations were higher for this approach than they were for stimulant or behavioral treatments. The hope that the cognitive-behavioral approach was the long awaited panacea was, of course, totally unrealistic. It is also the case that behavioral and pharmacologic approaches preceded formal CBT programs and achieved some notable successes before CBT was a viable alternative. Stimulant therapies, for example, typically have at least circumscribed success with 60-90% of hyperactive children. Thus for many practitioners, CBT has a greater evaluative burden than do other approaches to hyperactivity: The evidence of efficacy must be incremental, that is, beyond the effects of conventional treatments. To date, the findings have not met this demand. (Abikoff, 1985; Brown, Borden, Wynne, Schleser & Clingerman,

1986). One of the most carefully conducted studies of CBT for hyperactive children who were already taking stimulant medication yielded the discouraged and discouraging conclusion that, "There was virtually no evidence of efficacy of cognitive training. No significant treatment effects were obtained on academic, behavioral, and cognitive measures" (Abikoff & Gittelman, 1985, p. 953). This 16-week treatment study is particularly noteworthy not only because of the care that went into its design and implementation, but also because of the wide range of techniques, contexts, and response domains included.

In summary, CBT, like other treatment approaches to hyperactivity, has had delimited success. It is not that CBT has proven inferior to its alternatives, but rather that greater differential efficacy was expected, and claims of such differential effects rest, so far, on a rickety empirical base. CBT also requires more time, planning, and commitment than do pharmacologic and many behavioral approaches, and these additional costs must be included in any decision matrix.

A Realistic View and Reasons for Optimism

Despite the uneven track record of CBT, and the labor-intensiveness of the approach, there are several reasons to continue to apply and test cognitive strategies. One is that definitive empirical tests have not yet been conducted on a scale of sufficient breadth and depth. Methodological difficulties abound, and logistic and ethical complexities further constrain evaluation attempts. The failure of programs that employ two or three CBT sessions per week over a 6-or 12-week course to effect marked improvement is hardly sufficient reason to discard an entire approach to the treatment of multiproblem children.

The potential of CBT is not as the exclusive treatment of choice for ADD, but rather as an adjunct to other therapeutic approaches, included to achieve specific rather than global goals. When combined with tutoring to buttress basic skills, CBT can help the child acquire a set of problem solving strategies and learn how to match strategy to task and situation. For children who respond too readily and repetitively, CBT can encourage delay, teaching them to inhibit their immediate inclinations long enough to plan, evaluate, and revise. The CBT focus on conceptualization can also help children incorporate the changes they experience into their own theories of behavior and their own expectancies about future outcomes, rather than merely reacting reflexively to external manipulations. To the extent that the child's competencies are increasing at the same time he is encountering the message of self-control, CBT may prevent the development of maladaptive self-attributions suggested by pharmacologic or behavioral treatments. CBT may also facilitate drug discontinuance by buttressing the child's self-perceived efficacy and decreasing the importance that he and his significant others attribute to medication.

At this point, CBT is more an attitude or perspective—a style of administering treatment—than a bag of techniques. This attitude can have salutary effects in itself that are not measured in typical outcome studies. It also appears that CBT has proven beneficial by encouraging clinicians to assume a "cognitive stance" with more conventional treatments. Even when specific cognitive strategies are not included in the treatment program, there is often an increased recognition in contemporary programs of the importance of the cognitive context of therapeutic interventions. What has been criticized previously as a mindless behaviorism is more frequently found to be mindful since the advent of CBT. Contingencies, medications, and other external applications are no longer perceived as having homogeneous and unidimensional effects, but rather as having varying influences, depending upon how a child and his parents, peers, and teachers construe such treatments. There is growing recognition that treatments have inadvertent message values, and that such messages may either enhance or impede long-term goal attainment, even when short-term objectives are achieved.

Another important contribution of CBT is the portrayal of the child as an active agent who is not only competent to participate in his own treatment, but whose participation often aids the therapeutic enterprise. The child becomes the co-producer of therapeutic progress, conducting, in conjunction with the therapist, various experiments to illuminate and enhance his own actions, thoughts, and feelings. Rather than being a passive receptacle of CBT, the child collaborates with his therapist, helping to identify problem areas, design intervention strategies, evaluate changes, and return to the drawing board when progress is stayed. In summary, a major ingredient is the agentic posture—the view of the child as agent. This notion is integral to CBT but can also be incorporated into other treatment approaches.

Dilemmas for the Scientist-Practitioner

Although much more realistic than earlier notions about CBT, these potential benefits can still be construed as mere promissory notes. The basic compatibility and face validity of CBT, its humanistic premises, and some empirical successes all justify continued exploration and evaluation, as long as alternative treatments are not withheld. The justification for such continued exploration, however, is not scientifically based. It may be the case that much of the value of CBT cannot be subjected readily to systematic evaluation at this time. We are not yet skilled at measuring, for example, a child's perceived personal causation, his willingness to confront new challenges, and his notion of regulation of the self as a strategic act.

This is not one of those instances that allows a gracious exit following the caveat, "Further research is needed." Additional studies should and will be conducted, of course, but they may not provide conclusive answers on a global scale. Such studies will be valuable to the extent that they help develop and hone useful techniques and specify the conditions under which individual strategies are effective, futile, or even harmful.

In summary, the hope for CBT is derived more from the stance than from the data. The gap between expectations and evidence strains the tolerance of any respectable scientist, yet the fundamental premises continue to entice and encourage. This is the basic dilemma for the scientist-practitioner, but it is also the challenge that should spur further exploration and creative applications of this promising approach to treatment. To facilitate these efforts, we end with a set of guidelines and recommendations for the use of CBT with ADD.

COGNITIVE STRATEGIES FOR COGNITIVE-BEHAVIOR THERAPISTS: SUGGESTIONS AND GUIDELINES

1. CBT should be viewed as an adjunct rather than a replacement for other treatments—a useful addition to the therapeutic armamentarium. In many cases, stimulant medication, conventional behavioral methods, parent counseling, and academic tutoring are indicated for optimal effect.
2. CBT will not be beneficial in all instances, and care should be

taken to assess relevant characteristics of the child, family, and situation and evaluate early sessions before a major program is launched.
3. CBT should be individualized, taking into account each child's specific difficulties and the cognitive-behavioral processes that pose the greatest problems. There is probably no generic self-regulation deficit, but rather a large family of component processes that may create difficulties for any particular child, sometimes singly and other times in combination, perhaps in isolated contexts, perhaps across tasks and situations.
4. CBT should also be tailored to individual differences in temperament, learning styles, responsiveness to reinforcement, and other personal characteristics.
5. Teaching children how to get started and to sustain their own interests may be as important as teaching them problem-solving strategies.
6. Teaching children to anticipate and cope with delays and failure may be as important as teaching them techniques to facilitate success.
7. Generalization should be programmed rather than presumed, and global effectiveness should not be expected. Variability is the key here—in tasks, settings, trainers, methods, etc. It may also be helpful to train parents, siblings, teachers, peers, and others who interact with the child in various natural environments, in order to encourage the application of these techniques in as many settings as appropriate and possible.
8. Discrimination training may be as important as generalization training. Children need to learn not only how to use cognitive problem-solving strategies, but also when to use them, and when they may be inappropriate or counterproductive.
9. The timing of interventions may prove critical. Perhaps CBT should be initiated before stimulant treatment so that children will attribute gains to their own developing competencies rather than to a magic pill (Bugental, Whalen & Henker, 1977).
10. The timing of outcome evaluations may be equally important. Different treatments may prove effective for different problem domains, and they may exert their effects according to

different timetables. CBT may, for example, require more time to "take" than behavioral or pharmacologic approaches. This means that multiple rather than single-occasion assessments are needed.
11. Long-term maintenance cannot be assumed. Careful consideration should be given to the feasibility of "booster sessions," and follow-up evaluations are critical.
12. While focusing on cognitions and behaviors, it is important *not* to ignore emotional patterns. Many impulsive and hyperactive children have serious difficulties recognizing, regulating, and responding to their own affective reactions. Direct training in the complex interplay among thoughts, feelings, and overt actions may be required (e.g., Hinshaw, Henker & Whalen, 1984b).
13. Situational features are often more readily and more appropriately modified than personal characteristics. Rapid behavioral improvements may occur following relatively simple restructuring of a classroom setting or a family routine. The relevance of such situational dimensions may be obscured by an exclusive person orientation to treatment.
14. Analogously, it is important to teach children that behavior change on their part is not always required or desirable. They need to learn environmental programming as well as self-programming so that they will be able to assess, select, and modify their environments to optimize their performance.
15. Attend to children's self-attributions as well as to their self-instructions, to assess whether dysfunctional causal reasoning may be impeding progress or creating new difficulties.
16. The message of personal responsibility that is implicit in CBT should be communicated in synchrony with the child's developing competencies in an effort to minimize vulnerability to counterproductive guilt and blame.
17. The cognitive-behavioral therapist should be alert to the possibility of harmful or intraogenic effects rather than assuming that CBT is, by definition, benign.
18. The implicit message values of all treatments should be recognized so that attributional training can accompany other procedures when indicated, in order to prevent children from assuming either too much or too little responsibility for their behaviors and outcomes.

19. Treatment palatability and consumer satisfaction should be assessed along with behavior change, particularly when the intervention is demanding and lengthy.
20. Creative as well as standard procedures and combinations should be explored. One example is to encourage children to use newly acquired "cognitive prosthetics" to design a method to help them withdraw from medication. Another is to recruit children as co-therapists for a peer or younger child, thus affording them the opportunity to practice, test, and consolidate their developing self-regulatory skills.

NOTE

1. The acronym ADD is used to refer to attention deficit disorder with or without hyperactivity.

REFERENCES

Abikoff, H. (1985). Efficacy of cognitive training interventions in hyperactive children: A critical review. *Clinical Psychology Review, 5*, 479-512.

Abikoff, H., & Gittelman, R. (1985). Hyperactive children treated with stimulants: Is cognitive training a useful adjunct? *Archives of General Psychiatry, 42*, 953-961.

Brown, R. T., Borden, K. A., Wynne, M. E., Schleser, R., & Clingerman, S. R. (1986). Methylphenidate and cognitive therapy with ADD children: A methodological reconsideration. *Journal of Abnormal Child Psychology, 14*, 481-497.

Bugental, D. B., Whalen, C. K., & Henker, B. (1977). Causal attributions of hyperactive children and motivational assumptions of two behavior change approaches: Evidence for an interactionist position. *Child Development, 48*, 874-884.

Dodge, K. A., & Frame, C. L. (1982). Social cognitive biases and deficits in aggressive boys. *Child Development, 53*, 620-635.

Douglas, V. I. (1983). Attentional and cognitive problems. In M. Rutter (Ed.), *Developmental neuropsychiatry* (pp. 280-329). New York: Guilford Press.

Försterling, F. (1985). Attributional retraining: A review. *Psychological Bulletin, 98*, 495-512.

Henker, B., & Whalen, C. K. (1980). The many messages of medication: Hyperactive children's perceptions and attributions. In S. Salzinger, J. Antrobus, & J. Glick (Eds.), *The ecosystem of the "sick" child: Implications for classification and intervention* (pp. 141-166). New York: Academic Press.

Henker, B., Whalen, C. K., & Hinshaw, S. P. (1980). The attributional contexts of cognitive intervention strategies. *Exceptional Education Quarterly, 1*, 17-30.

Hinshaw, S. P., Henker, B., & Whalen, C. K. (1984a). Cognitive-behavioral and pharmacologic interventions for hyperactive boys: Comparative and combined effects. *Journal of Consulting and Clinical Psychology, 52*, 739-749.

Hinshaw, S. P., Henker, B., & Whalen, C. K. (1984a). Self-control in hyperactive boys in anger-inducing situations: Effects of cognitive-behavioral training and methylphenidate. *Journal of Abnormal Child Psychology, 12*, 55-77.

Kendall, P. C. (1985). Toward a cognitive-behavioral model of child psychopathology and a critique of related interventions. *Journal of Abnormal Child Psychology, 13*, 357-372.

Kendall, P. C., & Braswell, L. (1985). *Cognitive-behavioral therapy for impulsive children.* New York: Guilford Press.

Mahoney, M. J. (1977). Reflections on the cognitive-learning trend in psychotherapy. *American Psychologist, 32*, 5-13.

Mahoney, M. J., & Nezworski, M. T. (1985). Cognitive-behavioral approaches to children's problems. *Journal of Abnormal Child Psychology, 13*, 467-476.

Meichenbaum, D. (1985, June). *Cognitive behavioral modification with hyperactive children.* Paper presented at the International Congress on Hyperactivity as a Scientific Challenge, Gronigen, The Netherlands.

Meichenbaum, D. H., & Goodman, J. (1971). Training impulsive children to talk to themselves: A means of developing self-control. *Journal of Abnormal Psychology, 77*, 115-126.

Meyers, A. W., & Craighead, W. E. (Eds.). (1984). *Cognitive behavior therapy with children.* New York: Plenum Press.

Rosén, L. A., O'Leary, S. G., & Conway, G. (1985). The withdrawal of stimulant medication for hyperactivity: Overcoming detrimental attributions. *Behavior Therapy, 16*, 538-544.

Tharp, R. G., & Gallimore, R. (1985). The logical status of metacognitive training. *Journal of Abnormal Child Psychology, 13*, 455-466.

Whalen, C. K., Henker, B., & Hinshaw, S. P. (1985). Cognitive-behavioral therapies for hyperactive children: Premises, problems, and prospects. *Journal of Abnormal Child Psychology, 13*, 391-410.

What Is the Role of Group Parent Training in the Treatment of ADD Children?

Russell A. Barkley, PhD

ABSTRACT. The nature of the behavioral deficiencies in children with Attention Deficit Disorders and how these might lead to the development of noncompliance, oppositional behavior, and other conduct problems are briefly described. Since the majority of ADD children eventually display such conduct problems, it is important to train parents in behavior management skills so as to effectively treat these conduct problems. The abundance of ADD children referred to child guidance centers and the significant distress they create for their parents argue for the use of group parent training in child management skills as the most cost-effective method for managing this heavy case load. The rationale for a group parent training program, and the steps involved in it are briefly described. A case example is provided.

Young children with Attention Deficit Disorder (ADD) are plagued with numerous behavioral deficits, chief among these being inattention, impulsivity, restlessness, and poor self-control. These cognitive handicaps emanate into social interactions, resulting in frequent and significant interaction conflicts with others. Research in our clinic on parent-child interactions of ADD children routinely finds that such children are less compliant, more defiant and hostile, and less responsive to the interactions of their parents than are normal children. The parents of ADD children typically give more commands, directions, and general supervision, are less attentive to and rewarding of play and compliance, and reprimand and threaten punishment more often than parents of same-age normal children (Barkley, 1985). These interaction difficulties are also observed in the interactions of teachers and peers with ADD children (Mash &

Russell A. Barkley is Director of Psychology, Department of Psychiatry, University of Massachusetts Medical Center, 55 Lake Avenue North, Worcester, MA 01605.

© 1987 by The Haworth Press, Inc. All rights reserved.

Barkley, 1986; Pelham & Bender, 1982). Such interaction patterns, I believe, are the precipitants for referral of these children for treatment in child guidance clinics and constitute the most common complaints by their parents during initial evaluations. Evidence to date suggests that these cognitive and social deficits of ADD children are likely to persist into adolescence and young adulthood.

Recent research also suggests that mothers of ADD children are likely to report higher levels of stress and conflict in their role as parent of the child, greater depression, and more marital discord than families of normal children (Befera & Barkley, 1985; Mash & Johnston, 1983). Coupled with the frequent day-to-day conflicts in child management experienced by these parents, it is clear that psychological interventions must address the interaction problems, provide parents with more effective behavior management skills, and prepare them for the chronicity of the child's cognitive handicapping condition.

CLINICAL CONCEPTUALIZATION OF ADD CHILDREN AND THEIR FAMILIES

I believe that the deficits in sustained attention, impulse control, regulation of activity levels, and adherence to rules in controlling behavior (self-control) are handicaps in the development of these neuropsychological functions in ADD children. Such handicaps are presumed to have a neurophysiological basis, albeit as yet unspecified, which may prove to be either inherited or acquired early in development. As yet, the neurological substrates involved cannot be corrected by present interventions, although stimulant medications may provide temporary and partial amelioration of the problem at this level in many children so long as children remain on medication. In short, such drugs provide only partial symptomatic relief.

These cognitive handicaps result in frequent and serious social problems for ADD children, as they impair the children's ability to meet age-appropriate situational demands made by caregivers (parents, teachers, etc.) and peers. Hence, problems will often arise in compliance with commands, directions, rules, and social etiquette in the home, in academic performance and proper decorum in the classroom, in cooperative play and activities with peers, and in acceptable conduct of ADD children in the community. As a result of their incapacity to sustain attention and compliance to instructions, they will receive a disproportionate share of reprimands, censure, punishment, teasing, and ostracism from others, often beginning very early in development (ages 2 to 3 years). This would make ADD children far more likely to acquire and display behaviors which allow them to escape or avoid situations which make excessive demands on their limited attentional and self-control capabilities. Whining, temper tantrums, ignoring, refusal, defiance, hostility, destructiveness, and verbal and physical aggression may increase as the ADD children learn that such "coercive" behaviors succeed in the immediate withdrawal by caregivers of such demands on the children, and the eventual reduction by the parents in even initiating these demands with ADD children. A corollary of this is that the degree to which a command or instruction requires sustained effort to a relatively boring, effortful, and unrewarding task, the greater will be the children's defiant and oppositional behavior.

The extent to which these aggressive, defiant, and oppositional behaviors are acquired, performed and engrained in the children's behavioral repertoire is highly dependent on reactions they receive in the family and school environment, that is, their success or failure in functioning as escape/avoidance behaviors in demanding situations. There is little doubt that these coercive behaviors will find fertile soil in those families where parental psychiatric problems, marital discord, and poor child management techniques are common, or where other family members already display high rates of similar coercive (aggressive) behaviors towards each other (Patterson, 1982). These conditions create inconsistent parental responses to the ADD children's initial ventures at coercive behavior, leading to a family environment where such behaviors will meet with frequent success. Their increased display by the children over time and their eventual manifestation with other authorities outside the home seem an inevitable consequence. That such aggressive and defiant behaviors, if chronic, strongly predict a poorer adolescent and young adult outcome in ADD children has been well-established (Paternite & Loney, 1980; Thorley, 1984).

This conceptualizaton suggests that where parental management of the ADD children from an early age is consistent, immediate, and generally in accord with effective child management principles, where parental psychiatric problems are minimal or nonexistent, and where marital discord and intra-family displays of coercive behavior are close to normal levels, the rates of oppositional, defiant, hostile and coercive/aggressive behaviors by ADD children will be far lower than in families of ADD children where such problems are typical. In short, this view holds that the basic deficits in ADD children (inattention, impulsivity, restlessness) are biological in origin and unlikely to be permanently altered by interventions. The oppositional, defiant, and noncompliant behaviors often shown in these children are viewed, however, as learned and, hence, capable of great improvement through intervention.

WHY NOT DIRECTLY TREAT THE ADD CHILD?

If the initial difficulties are with the ADD children themselves, why not intervene directly with them and not the parents? While intuitively appealing, evidence to date suggests that such direct interventions are at best only partially effective (in the case of stimulant medications), and more typically ineffective (in the case of psychological treatments). There are a number of reasons why intervening with parents through group parent training may be far more useful:

1. Behavioral interventions applied by therapists directly to ADD children in clinic settings are highly unlikely to generalize to settings outside of the clinic involving other caregivers. Hence, training the adults responsible for the day-to-day care and management of the children permits those adults to become "therapists" who can have far greater and more frequent influence on the children than a clinician relying on once-per-week therapy sessions.
2. Individual treatment of ADD children fails to prepare parents adequately for the day-to-day interaction conflicts they will experience with these children, fails to provide information to the parents on the nature of the disorder, and fails to deal directly with their stress and self-confidence in the parental role, all of which are problematic with most families of ADD children.
3. Even where stimulant medication proves successful in improving the behavior of ADD children, parents will still require training in behavior management. These medications do not change all of the behavior problems of ADD children and are least likely to diminish the oppositional, defiant, and coercive behaviors or conduct problems. The time course over which the stimulants are effective, moreover, is quite brief, and by late afternoon, they are no longer exerting much influence over child behavior. Parents must therefore be prepared to deal with the unmedicated child in the evenings and on other occasions, such as weekends, where medication may not be

used. Finally, at least 20 percent of ADD children will not respond positively to medication, leaving little choice but to employ psychological interventions.
4. The relative abundance of ADD among clinic-referred children precludes most child guidance centers from providing individual therapy to all children having this disorder. Training parents via group instruction over a 10- to 12-week period provides far greater economy and cost-effectiveness. Evidence also suggests that group training of parents may be equally as effective as individual training with a single family.

GOALS OF INTERVENTION

The group training of parents is intended to achieve a number of purposes. First, parents must be provided with current knowledge on the nature of this disorder, its developmental course, and known etiologies. Such information is essential to counteract the plethora of misinformation available on the disorder in the popular media. It also serves a second purpose, which is to help parents come to realize that ADD is a developmentally handicapping condition for which, at present, there is no cure. Instead, an attitude of coping must be instilled that assists parents in lowering expectations for the child's behavior and performance, and altering the manner in which the child is treated so as to maximize the child's adjustment as much as possible. Third, training is intended to provide parents with behavior management skills that have previously proven effective in lessening the disruptive and noncompliant behavior of these children while increasing appropriate, prosocial behavior. This will hopefully achieve a fourth intention, that being the reduction of parental distress, coupled with feelings of increased competence in their role as parent and caretaker.

A Parent Training Program for Families of ADD Children

There are 10 steps in our program for parents of ADD children. These are briefly outlined below, and further information on them can be obtained from the text by Barkley (1981):

1. Providing parents with information about ADD. Parents receive detailed information on the nature, associated

features, etiologies, developmental course, outcome, and proven and disproven treatments for ADD.
2. Instructing parents in the processes and concepts involved in child misbehavior and the reasons for its development and maintenance. Parents are taught that misbehavior stems from four levels of causes, each interacting with the others to culminate in behavioral problems. These levels are child characteristics and competencies (e.g., developmental delays, negative temperament, health problems, etc.), parent characteristics and competencies, consequences for misbehavior when it occurs (e.g., escape or avoidance of chores and work, increased parental attention, etc.), and family setting or stress events (e.g., marital discord, parental unemployment or financial problems, problems with siblings or other family members, etc.).
3. Training parents in ways to develop and enhance their positive attention to the ADD children and their appropriate behavior. Such parents often rely heavily on reprimands, punishment, and other aversive behaviors to manage child misbehavior, but they may display limited amounts of positive attention to appropriate child behavior. This step attempts to teach parents to differentially attend (positively) to ongoing child behaviors that are appropriate and nondisruptive, while ignoring minor child misbehavior.
4. Extending parental attention to child compliance. This step then increases the use of parental positive attention for reinforcing child compliance with commands and rules of the household.
5. Using home token reinforcement systems. It is naive to assume that the use of parental praise and attention alone will substantially alter misbehavior in ADD children. Such children appear to require more frequent and salient consequences if their appropriate behavior is to be sustained. To achieve this end, parents are trained in the implementation of a "poker chip program" or home point system for reinforcing the children's performance of chores and other required activities.
6. Employing response cost and time out as disciplinary tactics. Parents now receive training in the use of fines within the home token systems for minor rule infractions and noncompliance. Subsequently, they receive intense instruction in a highly effective time-out technique which relies on immedi-

ate isolation of the children to a dull corner of the home contingent on noncompliance.
7. Extending time out to other home misbehavior. Since parents are restricted to using time-out for only one or two types of misbehavior in the previous step, this step focuses on troubleshooting problems that are experienced by the parents in the prior step, and then extending the use of time out to several other misbehaviors by the children.
8. Managing child misbehavior in public places. Prior to this step, parents are requested not to use the techniques to manage behavior problems in stores, restaurants, church, or other public places. The nature of these settings is such that minor modifications to the use of the management methods is required to employ them effectively in public places. The intent of this step is to discuss in detail how prior management methods can be effectively used in these settings. Parents are further instructed in a "think aloud-think ahead" procedure wherein, prior to entering a public place, a plan for managing misbehavior is developed and shared with the children.
9. Managing future misbehavior. This step challenges parents with types of misbehavior not yet shown by the children and requires them to design a method of management for that problem based upon the methods learned to date. It is hoped that this step encourages parents to generalize their new management skills to possible future problems they may encounter with their ADD children.
10. One-month booster session. Following training, parents are seen in one month to review possible problems developing in the management of the children, to review prior techniques and information as needed, and to render any further services which may be needed.

Case Illustration

Michael, a 5-year-old child, was initially evaluated because of problems with poor sustained attention, restlessness, impulsivity, and defiant/oppositional behavior. His parents, both in their early 20s in age, were distraught over their inability to get Michael to mind them, as well as the outright defiance he displayed toward their authority. Michael's mother was quite depressed over her perceived incompetence to deal with Michael, while his father had received treatment previously not only for depression, but for alcoholism as well. Michael's father was currently unemployed, adding further to the stress within the household. Both parents had at times requested the Department of Social Services to place Michael in foster care due to their inability to cope with his ADD and conduct problems. It was clear from the evalution that a number of problems existed with Michael and the family. These were delimited into four areas: Michael's ADD; the parent's inconsistency in managing Michael and heavy reliance on punishment; parental psychiatric problems; and stress due to the father's unemployment. Each area would require intervention if any success would be made in improving the family's present circumstances.

Following his evaluation and diagnosis as both ADD and Oppositional Disorder of Childhood, Michael was referred to his pediatrician for placement on Ritalin, a stimulant drug frequently used to treat ADD. Michael's father returned to an alcohol treatment program, joined Alcoholics Anonymous for supportive counseling, and began receiving individual therapy and medication for his depression. Michael's mother successfully obtained employment to ease the family's financial problems until his father could find work. Both parents were then enrolled in the aforementioned child management training program with four other families having ADD children. To date, substantial changes in Michael's behavior and parental management skills have been made. Medication has helped to diminish the inattention, restlessness, and impulsivity of this child, which decreases parental distress and permits greater use of positive reinforcement methods with Michael. Michael's father remains on antidepressant medication but no longer requires individual therapy. He has returned to job hunting and believes success in doing so is imminent. Michael's mother expresses much greater confidence in dealing with his misbehavior, when it occurs, and both parents have come to accept Michael as a behavioral handicapped child in need of acceptance, compassion, somewhat reduced expectations, and special management. Michael, in turn, has become more compliant with requests, and displays less defiance and oppositional behavior and fewer tantrums and destructiveness.

SUMMARY

It should be clear from this case that the role of any single treatment is often limited in its efficacy in treating the multiple problems families with ADD children often have. However, group parent training in child management skills has a critical role to play in preparing the parents for the many years of coping with behavioral problems that will be necessary to satisfactorily raise such children. When combined with medication, special education, and treatment of the parents' own personal or emotional problems as needed, success in addressing the many problems of these families can often be achieved. Nevertheless, periodic retraining of the parents through booster sessions, reinvolvement of professional help for new problems that will arise, and special educational services are often required in the long-term care of ADD children.

REFERENCES AND SUGGESTED READINGS

Barkley, R. A. (1981). *Hyperactive children: A handbook for diagnosis and treatment.* New York: Guilford Press.
Barkley, R. A. (1985). Family interaction patterns in hyperactive children. In D. Routh & L. Wolraich (Eds.) *Advances in behavioral pediatrics.* Greenwich, CT: JAI.
Barkley, R. A. (1987). *Training parents to manage behavior problem children.* New York: Guilford Press.
Befera, M. S., & Barkley, R. A. (1985). Hyperactive and normal boys and girls: Mother-child interaction, parent psychiatric status, and child psychopathology. *Journal of Child Psychology and Psychiatry, 26,* 439-452.
Bidder, R. T., Gray, O. P., & Newcomb, R. (1978). Behavioural treatment of hyperactive children. *Archives of Diseases of Children, 53,* 574-579.
Dubey, D. R., O'Leary, S. G., & Kaufman, K. F. (1983). Training parents of hyperactive children in child management: A comparative outcome study. *Journal of Abnormal Child Psychology, 11,* 229-246.
Mash, E. J., & Barkley, R. A. (1985). Assessment of family interactions with the Response Class Matrix: A review of research. In R. Prinz (Ed.), *Advances in behavioral assessment of children and families.* Greenwich, CT: JAI.
Mash, E. J., & Johnston, C. (1983). Parental perceptions of child behavior problems, parenting self-esteem, and mother's reported stress in younger and older hyperactive and normal children. *Journal of Consulting and Clinical Psychology, 51,* 68-99.
Paternite, C., & Loney, J. (1980). Childhood hyperkinesis: Relationship between symptomatology and home environment. In C. Whalen & B. Henker (Eds.), *Hyperactive children: The social ecology of identification and treatment.* New York: Academic Press.
Patterson, G. R. (1982). *Coercive family process.* Eugene, OR: Castalia.
Pelham, W. E., & Bender, M. E. (1982). Peer relationships in hyperactive children: Description and treatment. In K. Gadow (Ed.), *Advances in learning and behavioral disabilities.* Greenwich, CT: JAI.
Thorley, G. (1984). Review of follow-up and follow-back studies of childhood hyperactivity. *Psychological Bulletin, 96,* 116-132.
Willis, T. J., & Lovaas, I. (1977). A behavioral approach to treating hyperactive children: The parent's role. In J. G. Millichap (Ed.), *Learning disabilities and related disorders.* Chicago, IL: Yearbook Medical Publications.

What Is the Role of Academic Intervention in the Treatment of Hyperactive Children with Reading Disorders?

Ellis Richardson, PhD
Samuel Kupietz, PhD
Steven Maitinsky, MD

ABSTRACT. The clinical interaction of Attention Deficit Disorder with Hyperactivity (ADD/HA) and Developmental Reading Disorder (DRD) is discussed. A treatment program that is particularly suited to children with ADD/HA and DRD is described. A recently completed study of the effects of methylphenidate on the reading achievement of ADD/HA children is described and selected results are presented to support the conclusions that: (a) special reading instruction can substantially improve achievement rates in children with ADD/HA and DRD, and (b) the degree to which methylphenidate reduces the behavioral symptoms of ADD/HA is a crucial factor in determining a child's response to DRD therapy.

First, we discuss considerations in the interaction of Attention Deficit Disorder with Hyperactivity (ADD/HA) and Developmental Reading Disorder (DRD) and procedures for correcting DRD in the hope that our own experience will be of value to others who treat these children. In the second half of the chapter, we review results from a study that we recently completed which have important implications for the treatment of children with a dual diagnosis of DRD and ADD/HA.

Ellis Richardson and Samuel Kupietz are with the Nathan S. Kline Institute for Psychiatric Research, Orangeburg Road, Orangeburg, NY 10962. Steven Maitinsky is with the Child Development Center, Nassau County Medical Center, 2201 Hempstead Turnpike, East Meadow, NY 11554. This work was supported in part by a grant from the National Institute of Mental Health (#RD1 MH 36004).

© 1987 by The Haworth Press, Inc. All rights reserved.

INTERACTION OF ADD/HA WITH DRD

The interaction of ADD/HA and DRD is an extremely important consideration in treating children with a dual diagnosis. In some children, the ADD/HA symptoms may be contributing to the reading problem. In such cases, DRD may be diagnosed even when none of the positive indicators of DRD are present. Most children, however, who are diagnosed both ADD/HA and DRD do show positive signs on one or more of the DRD indicators. In either case, DRD and ADD/HA can interact in such a way that they exacerbate one another.

The child who manifests ADD/HA symptoms in the classroom is obviously at risk for reading failure. Frequent shifts of attention during instruction will interfere with the acquisition of basic skills, and poor classroom behavior may unwittingly (or wittingly) result in less active instructional time afforded that child (i.e., the teacher may simply avoid the thankless task of trying to hold the child's attention long enough to get the material across). Fluctuations in attention can also affect the child's ability to synthesize ideas during contextual reading, interfering with comprehension and the development of decoding skills.

Conversely, reading failure might be expected to increase or, in some cases, even produce behavioral symptoms of ADD/HA. DRD is an emotionally painful disorder that causes tremendous stress and embarrassment both at school and at home. A child who is under such pressure is likely to act out in the classroom, diverting attention from the business of reading instruction. Also, frequent shifts in attention remove the child, if only momentarily, from the offending situation. Thus, when a child is handicapped both by ADD/HA and DRD, the behavioral disorder may be more severe as a result of the DRD condition *and* the reading deficiency may be more severe as a result of the ADD/HA condition.

TREATING DRD AND ADD/HA IN THE CLINICAL SETTING

A Treatment Program for DRD

For the past 20 years, we have been involved in a research program to develop better procedures for correcting DRD. An instructional model, the Integrated Skills Method (ISM) (Richardson & Bradley, 1974), was first described over a decade ago and more recent procedural developments are presented by Richardson (1984). The ISM provides a method for the therapist to exercise modality-independent control over the treatment program (i.e., independent management of sight-word and phonic/linguistic instructional strategies). Thus, for example, when working with a child who learns sight-words rapidly but who has serious problems with phoneme processing, the therapist may move the child rapidly through a basal reader that stresses sight-words while conducting an intensive program to remediate the phonic/linguistic decoding skill deficit. Conversely, a child with different strengths and weaknesses may proceed through the phonic/linguistic skills sequence rapidly while the therapist structures other exercises to improve his/her ability to handle the basal reader.

A diagram describing the ISM is presented in Figure 1. It is generally recognized that words may be decoded (i.e., read) in two ways: The printed word may be directly associated with meaning, or it may be analyzed into graphemic units corresponding to phonemic sounds and which are then associated with meaning. This observation has given rise to two distinct methods for teaching reading: the sight-word approach and the phonics approch (Chall, 1967).

The ISM employs both types of instructional strategies. Sight-Word processing (Basal Vocabulary—e.g., *the*, *where*, *you*) teaches immediate recognition of words that are likely to be encountered at the child's reading level. Phonic/Linguistic processing teaches word recognition through the use of linguistic spelling patterns. For example, the *an* pattern enables a child to read *man*, *pan*, *fan*, etc.; the *ay* pattern enables the child to recognize such words as *may*, *bay*, *say*, and *day*, etc.; the *ight* pattern enables the child to recognize such words as *sight*, *right*, *bright*, and *light*.

More recent developments in the treatment program exploit the interaction of these two modes in contextual reading. That is, children are taught to process sound patterns while simultaneously searching for a correspondingly meaningful word that is appropriate in the context and they are taught to encode (i.e., spell and write) words according to linguistic patterns. We have observed surprisingly rapid gains in some children using these new techniques.

The Sight-Word curriculum is defined by the vocabulary sequence of any standard vocabulary-controlled basal reader pro-

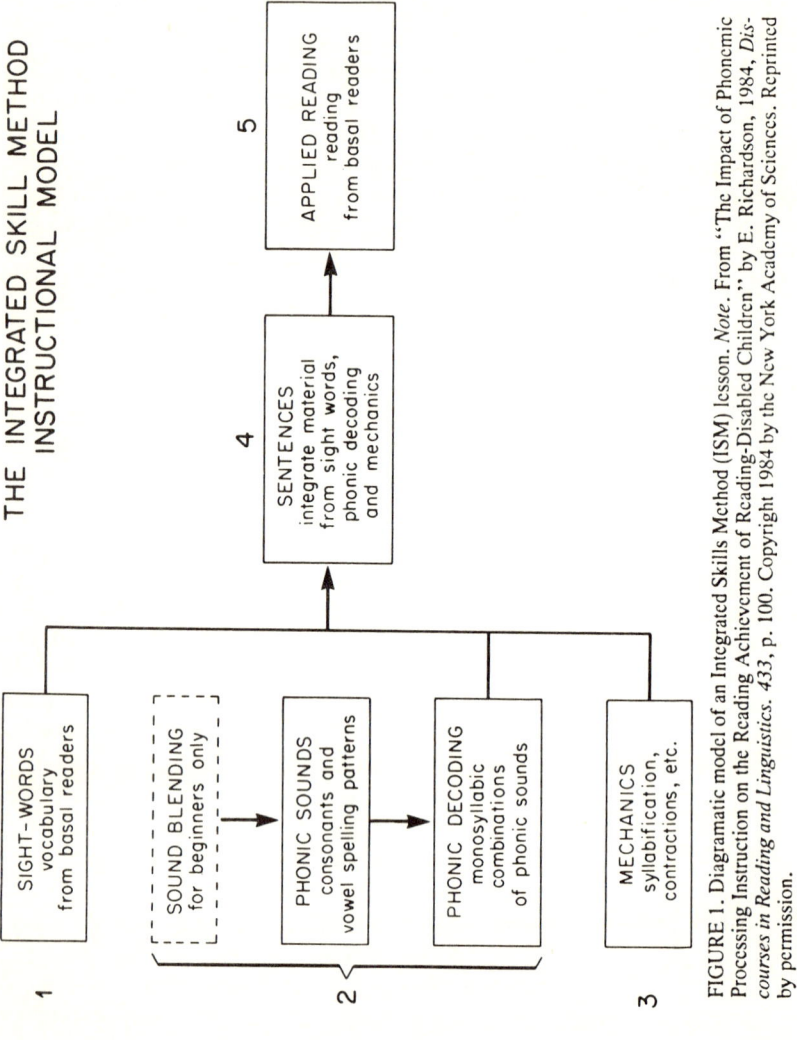

THE INTEGRATED SKILL METHOD
INSTRUCTIONAL MODEL

FIGURE 1. Diagramatic model of an Integrated Skills Method (ISM) lesson. *Note.* From "The Impact of Phonemic Processing Instruction on the Reading Achievement of Reading-Disabled Children" by E. Richardson, 1984, *Discourses in Reading and Linguistics. 433*, p. 100. Copyright 1984 by the New York Academy of Sciences. Reprinted by permission.

gram. We use the Macmillan Series "r" readers in our clinics, but the ISM has been used with many other standard reading programs. In the clinic, children are tested on words to be introduced in the basal reader stories and those words which the child does not recognize immediately are written on index cards for practice.

The Phonic/Linguistic processing component of the instructional model teaches children to use the letter-sound system to decode unknown words. Much of the research on the ISM has been devoted to this aspect of reading instruction (see, Richardson & DiBenedetto, 1977). In the first step, Phonic Sounds, the sounds represented by specific spelling patterns and consonants, are taught. This is followed by Phonic Decoding in which a child practices with a variety of words that are based on the given spelling patterns. (The Sound Blending step represented in Figure 1 is rarely needed with DRD children. These procedures are discussed in Richardson [1984].)

The third component, Mechanics, is used to teach a variety of material such as contractions, attention to sound patterns in common sight-words (e.g., knowing the word *look* may enable one to read *cook*, *book*, *took*, etc.), word endings (e.g., *ed*, *ing*, *er*), and the application of phonic skills to polysyllabic forms. For example, a child who has mastered short *a* and *i* patterns in Phonic Decoding can decode syllables such as *man*, *ban*, and *din* might be given items such as *manner*, *banner*, and *dinner* in the Mechanics step of the lesson.

The next step of an ISM lesson, Sentences, is used to teach the integration of basal vocabulary, phonic/linguistic, and mechanics skills during contextual reading. This also provides an opportunity for instruction in word meaning, comprehension and the interaction of these higher order skills with the more fundamental decoding skills. Implementation of this step is particularly difficult since it requires the therapist to write sentences that incorporate the material that is appropriate for each individual child.

The final component of the ISM, Applied Reading, primarily involves assisting the child in reading stories from basal readers. In sessions, the therapist helps the child integrate skills learned in prior lesson steps into the process of reading. Although the diagramatic model in Figure 1 emphasizes decoding skills, word meaning and language development are important aspects of all ISM lesson steps so that Applied Reading is primarily devoted to teaching the child to read for enjoyment and information.

Basically, the ISM involves no special techniques such as revised alphabets or colored vowels. Rather, it incorporates good standard instructional practices. The unique aspect of the ISM is its structure, which permits the therapist to exercise individualized control over the curriculum, the specific skills taught, and the shaping of more effective reading behavior in individual children.

The Therapeutic Setting

When working with ADD/HA children, the one-to-one setting is extremely important. Given the interactive proximity of the child and therapist, the therapist is able to insure sustained attention during reading. Fluctuations in attention observed by the therapist are signals to gently bring the child back to the task. Impulsivity during reading (children rushing through a given text, decoding without regard for meaning) can be controlled by slowing the child down and directing his/her attention to the meaning of the text. Finally, and most importantly, frequent verbal reinforcement during reading provides the child with the rare experience of receiving positive reinforcement for behavior which has heretofore been generally negatively reinforced. This improves the child's self-image and alleviates some of the pressures that exacerbate the ADD/HA symptoms.

Proper application of these procedures with children who have both ADD/HA and DRD should result in: (a) accelerated reading achievement, (b) improved attention during reading and better comprehension, and (c) a general reduction in ADD/HA behaviors.

"Proper application" is a key term in this clinical assertion. For example, it is extremely important that the material the child is asked to read be at a comfortable level. Such children have rarely had the opportunity to read material at a relatively easy level, and their introduction to a reader in which they do well sets the occasion for verbal reinforcement during reading. The therapist must be relatively experienced in on-going diagnosis of reading errors and how to handle them. For example, depending upon the child, the material being read, and the type of error made, a therapist may choose to respond to any given error by: (a) ignoring it completely, (b) correcting it immediately, (c) waiting until the child has finished a given portion of the text and then calling attention to the error, or (d) even preventing an anticipated error by prompting the child just before he/she gets to a particular word. The therapist must thoroughly understand the instructional model being applied and how it may be manipulated to insure maximum success for the child. We have intentionally avoided applying the term teacher here. Given the right circumstances, there are many teachers who are capable of applying the techniques discussed, but these skills are not necessarily a part of every teacher's repertoire. Since individualization is an important key to successful treatment, these procedures would not be expected to work for children with DRD and ADD/HA in a normal classroom setting. In the resource room, where three to five children may be handled at one time, a highly skilled teacher may be expected to improve the reading progress of many such children to some degree. For these procedures to work most effectively, however, a highly trained and skilled therapist is required. In the next section of this chapter we present one setting in which this model has been effectively applied.

A STUDY OF DRD AND ADD/HA

Study Description

We have just completed a five-year study of the effects of methylphenidate on the reading achievement of children with a dual diagnosis of ADD/HA and DRD. In that study, a complete data set was gathered on 42 children diagnosed as hyperactive, with this diagnosis confirmed by teacher ratings on Factor 4 (hyperactivity) of the Conners Behavior Rating Scale. The children were all between the ages of 7 and 12, had attained less than 75 percent of their expected reading level, and had WISC-R IQs of 85 or more on either the Verbal or Performance scales.

At pretest, the children were given the Gates-MacGinitie, the Peabody Individual Achievement Test (PIAT), and the Decoding Skills Test (DST). They were randomly assigned to one of three methylphenidate treatment groups: 0.3 mg/kg; 0.5 mg/kg; 0.7 mg/kg; or to placebo. They then received a total of 24 weekly ISM sessions and the pretest battery was repeated after 12 and 24 weeks of intervention. The children and parents were trained to use the ISM materials for daily practice at home between sessions.

The 39-item Conners Behavior Rating Scale was obtained from the children's school teachers at pretest (prior to the initiation of medication), two weeks later (following initiation of medication but prior to reading intervention), after three months of reading therapy, and at the end of the study (after 6 months of reading therapy).

Additional procedural details have been described by the first author (Richardson, 1984), and a complete presentation of the study's results is being prepared for publication. Here we present the results of several analyses which have a direct bearing on the purpose of this chapter, the interaction of DRD and ADD/HA and the implications for academic intervention.

Response to DRD Treatment

Figure 2 depicts the summarized reading scores of the children at pretest and after 3 and 6 months of treatment. The horizontal axis shows Study Week,[1] while the vertical axis indicates Reading Grade Equivalent (RGE) scores (obtained by averaging grade equivalent scores from PIAT Reading Recognition and Comprehension, Gates Vocabulary and Comprehension, and DST Basal Vocabulary and Contextual Reading subtests). The figure shows that the average RGE of these children was about 2.6 at pretest, while their average school grade placement (not shown in the figure) was 4.5, indicating that they were about two years behind in reading. By Study Week 14, they scored 3.2 on the RGE, and, by the completion of the six-month treatment program, they scored just above 3.5. Thus, during the course of the six-month study they had gained approximately one year in reading, reducing their reading deficit by about 4 months.

Unfortunately, since this study was designed to assess the effects of pharmacotherapy, no control for the reading program was included and, therefore, no conclusions regarding the specific effects of the reading program independently of drug treatment can be drawn. To obtain some estimate of the effects of the study program on the children's reading achievement, however, dependent *t*-tests were used to compare observed RGE scores at Study Weeks 14 and 28 with scores predicted from pretest and those predicted on the basis of normal achievement (see Figure 2). Results of this analysis revealed significant effects ($p < .01$) at both Study Weeks 14 and 28. Reference to Figure 2 shows that at Week 28, following treatment, the children were scoring about 6 to 7 months higher than predictions based on pretest scores and about 2 to 3 months higher than predictions based on a normal achievement rate.

The assessment of gains over normal resulting from academic

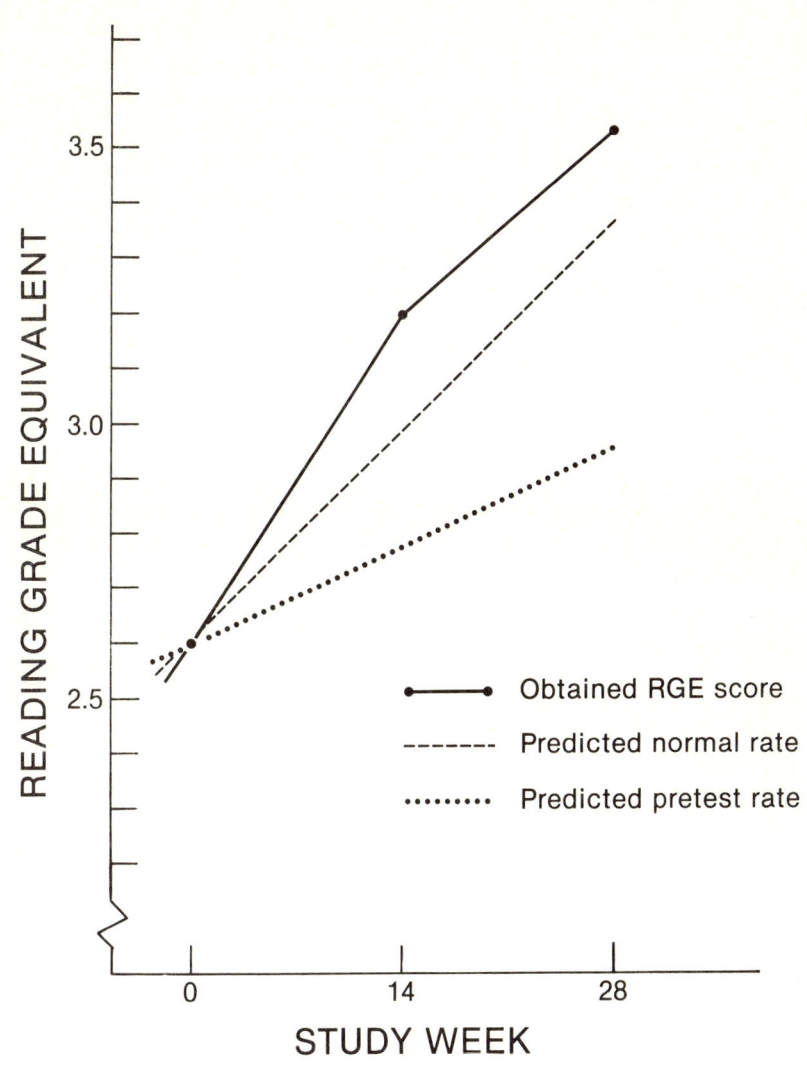

FIGURE 2. Mean Reading Grade Equivalent (RGE) scores as a function of Study Week at pretest (Week 0), mid-study (Week 14), and posttest (Week 28).

treatment is crucial since it reflects the degree to which the participating students are catching up. The children in this study were nearly 2 years behind in reading at the beginning of the study but, by the end of the study, this margin had been reduced to about 1.6 years. The positive aspects of this result must be qualified by the fact that, at this pace, it might take from 5 to 6 years of such treatment to get them "on grade level." This emphasizes the extreme importance of early identification and treatment for children with DRD. If they can be reached during first or second grades, before an extremely severe deficit has developed, it may be possible to correct the problem within a year. If diagnosis and treatment is postponed several years until a two- or three-year discrepancy develops, however, it may be virtually impossible to completely ameliorate the problem. Nevertheless, our results demonstrate that the achievement rates of ADD/HA children with DRD can be improved substantially through special reading instruction.

Response to Medication

The principal finding of the methylphenidate study is of particular relevance to this chapter. Methylphenidate treatment was shown to have a positive effect on reading achievement. The results, however, indicated that this effect was mediated by the degree to which medication induced behavioral changes as measured by the Conners Behavior Rating Scale scores. To demonstrate the importance of this effect, the 12 children whose Hyperactivity factor scores were most reduced by medication (Good Response group) and the 12 whose scores were reduced the least (Poor Response group) were selected from among the 34 children who actually received medication. Selections were based on observed change on the Conners Hyperactivity factor following two weeks of methylphenidate treatment but prior to the initiation of reading therapy. The mean Hyperactivity ratings obtained throughout the study for these two groups are shown in Figure 3.

Reference to Figure 3 reveals that prior to medication, at Study Week 0, both groups obtained mean Hyperactivity ratings above 3.0.[2] After two weeks of methylphenidate treatment, the mean Hyperactivity score of the Good Response group fell to nearly 1.5 and remained below 1.9 at subsequent measurement points. In contrast, the mean score for the Poor Response group decreased only slightly from about 3.0 to about 2.5 and remained there throughout the study. T-test comparisons revealed no significant pretreatment differences between the two groups on hyperactivity ratings, reading scores, age, or IQ. The differences in hyperactivity ratings at all three treatment points (Weeks 2, 14, and 28 in Figure 3), however, were all significant ($p < .05$).

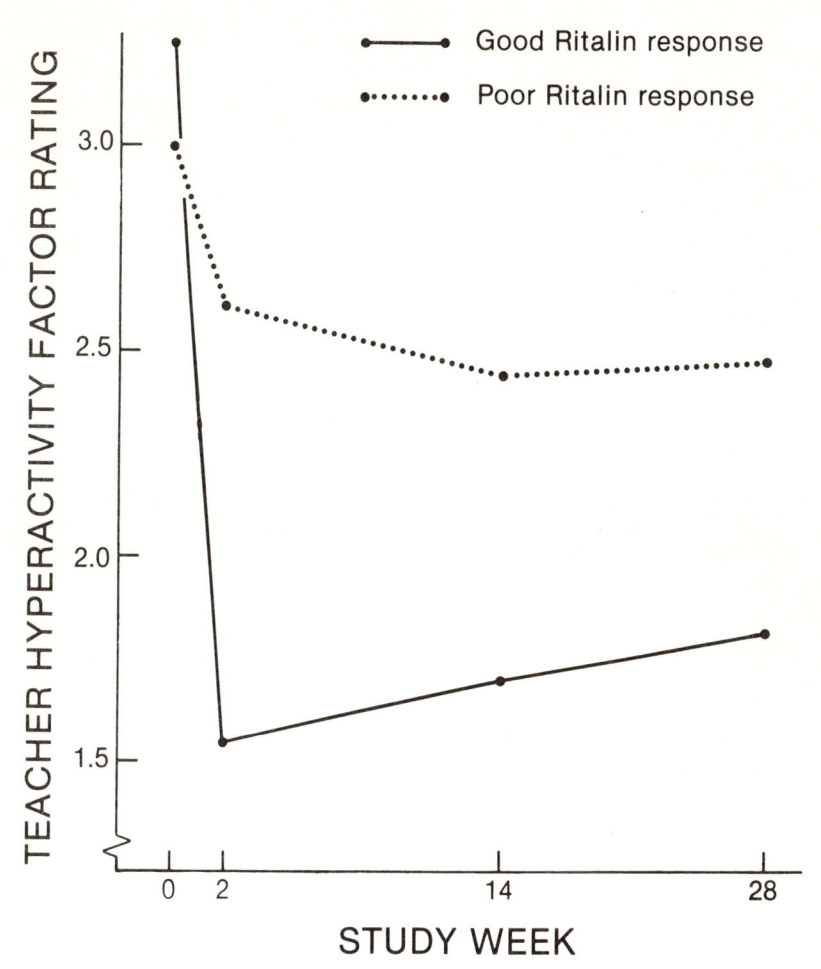

FIGURE 3. Mean Conners' Hyperactivity factor scores of Good and Poor Ritalin Response Groups as a function of Study Week.

To demonstrate how behavioral change affected reading achievement, mean RGE scores were computed for the Good and Poor Methylphenidate Response groups separately (see Figure 4). While the Good Response group scored only a little more than one month higher than the Poor Response group at pretest ($t = 0.49$), by Week 14 there was nearly a 7-month difference in the reading levels of the two groups ($t = 2.15$, $p < .05$) which was maintained at Week 28 ($t = 2.11$, $p < .05$).

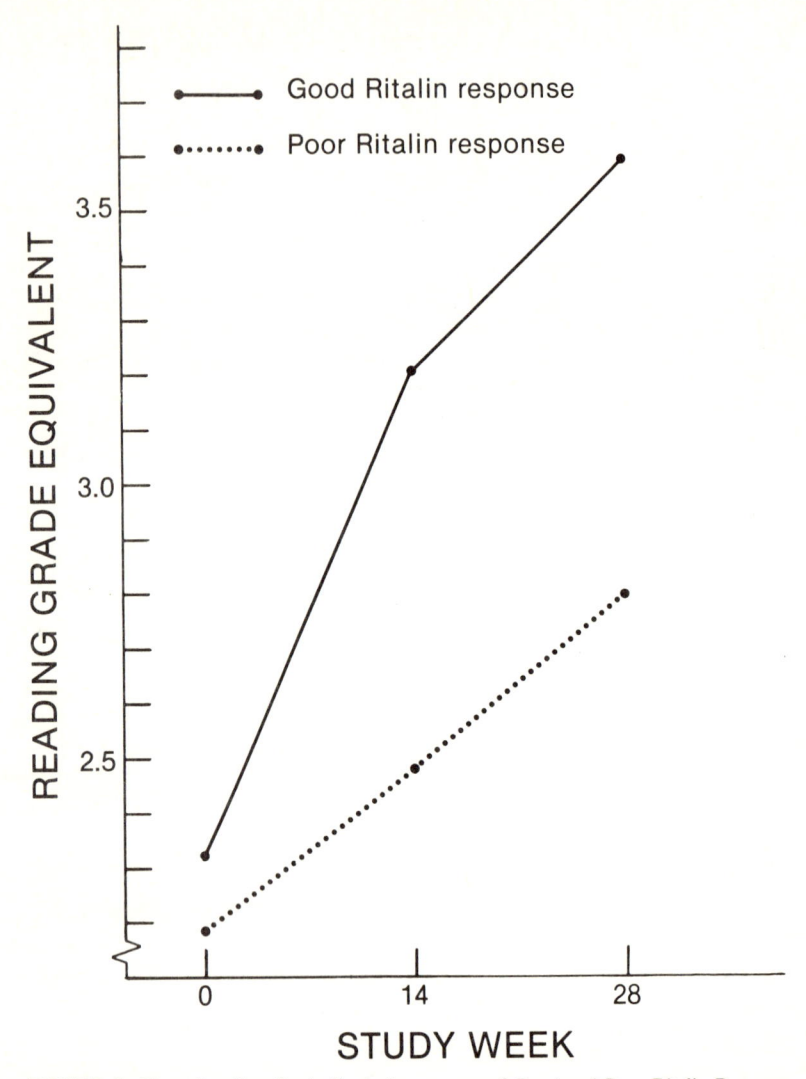

FIGURE 4. Mean Reading Grade Equivalent scores of Good and Poor Ritalin Response groups as a function of Study Week.

Although one must exercise caution in interpreting results based on grade equivalent scores, the magnitude of the differences shown in Figure 4 leaves little doubt that the effective treatment of ADD/HA behavioral symptoms can produce a dramatic difference in a child's response to reading instruction. Furthermore, these results provide strong support for our earlier clinical assertion regarding the interaction of ADD/HA and DRD. If a child's reading achievement rate can be changed radically by reducing ADD/HA symptoms, it follows that the ADD/HA condition must have been a potent factor in the DRD condition. These results open a number of clinical possibilities. Can other forms of therapy that reduce ADD/HA symptoms have a similar effect on DRD? To what degree does the effective treatment of DRD ameliorate ADD/HA symptoms? Would similar results be obtained in the absence of special reading instruction? Answers to these questions will have an immediate impact on the quality of clinical care afforded children with a dual diagnosis of ADD/HA and DRD.

SUMMARY AND CONCLUSIONS

Although the exact nature of the relationship has yet to be determined, it is likely that a functional relationship exists between DRD and ADD/HA. At present, efforts to study this relationship are impeded by the failure to develop positive diagnostic criteria for defining DRD. Research indicates, however, that phoneme processing problems, automatic verbal retrieval and sequencing, and language development problems are symptomatic of DRD. It is apparent from our own clinical research that most children with a compound diagnosis of ADD/HA and DRD can be taught effectively to read. The research results reviewed above demonstrate this quite clearly. Furthermore, the results demonstrate the interaction of ADD/HA symptoms with the treatment of DRD. The degree of success of the DRD treatment program was highly dependent upon the degree to which Methylphenidate alleviated the ADD/HA symptoms. Based on these results and our clinical experience with hyperactive and reading disordered children, we offer the following recommendations:

1. DRD treatment for these children should emphasize success.
2. DRD treatment should be based on a highly structured approach that stresses basic decoding skills while fostering higher-order processing skills.
3. DRD treatment should be based on individualized planning of lesson materials and should be offered in a one-to-one setting as frequently as possible (3 to 5 days per week).
4. DRD treatment should be combined with a carefully managed program of treatment for the ADD/HA.

NOTES

1. According to the study design, reading scores and behavior ratings were obtained prior to medical treatment (at Week 0). Two weeks later, at the initiation of methylphenidate and reading treatment (Week 2), behavior ratings were obtained a second time. Both reading and behavior ratings were obtained at Study Week 14 (following 3 months of reading instruction) and again at Study Week 28 (one week after the completion of the six-month reading program).
2. The Conners was scored by averaging responses to the items in each factor. Ratings on the items were: 4, "Very much," 3, "Quite a Bit," 2, "Just a Little," and 1, "Not at All."

REFERENCES

American Psychiatric Association. (1980). *Diagnostic and statistical manual of mental disorders* (3rd ed.) Washington, DC: Author.
Bradley, C. (1937). The behavior of children receiving Benzadrine. *American Journal of Psychiatry, 94*, 577-585.
Bradley, L., & Bryant, P.E. (1978). Difficulties in auditory organization as a possible cause of reading backwardness. *Nature, 271*, 746-747.
Chall, J. (1967). *Learning to read: The great debate.* New York: McGraw-Hill.
Chall, J., Roswell, F. G., & Blumenthal, S. H. (1963). Auditory blending ability: A factor in success in beginning reading. *Reading Teacher, 17*, 113-118.
Cantwell, D.P., & Satterfield, J.H. (1978). The prevalence of academic underachievement in hyperactive children. *Journal of Pediatric Psychology, 24*, 161-171.
Denckla, M.B., Rudel, R.G., & Broman, M. (1981). Tests which discriminate between dyslexic and other learning-disabled boys. *Brain and Language, 13*, 118-129.
DiBenedetto, B., Richardson, E., & Kochnower, J. (1983). Vowel generalizations in normal and learning disabled readers. *Journal of Educational Psychology, 75*, 576-582.
Dykman, R.A., Ackerman, P.T., & Holcomb, P.J. (1985). Reading disabled and ADD children: Similarities and Differences. In D. Gray and J. Kavanagh (Eds.), *Biobehavioral measures of dyslexia.* Parkton, MD: York Press.
Fox, B., & Routh, D.K. (1976). Phonemic analysis and synthesis as word-attack skills. *Journal of Educational Psychology, 68*, 70-74.
Halpern, J.M., Gittelman, R., Klein, D.F., & Rudel, R.G. (1984). Reading disabled hyperactive children: A distinct subgroup of Attention Deficit Disorder with Hyperactivity? *Journal of Abnormal Child Psychology, 12*, 1-14.
Kochnower, J., Richardson, E., & DiBenedetto, B. (1983). A comparison of the phonic decoding ability of normal and learning disabled children. *Journal of Learning Disabilities, 16*, 348-351.
Lambert, N.M., & Sandoval, J. (1980). The prevalence of learning disabilities in a sample of children considered hyperactive. *Journal of Abnormal Child Psychology, 8*, 33-50.
Liberman, I.Y., Shankweiler, D., Liberman, A.M., Fowler, C., & Fischer, F.W. (1977). Phonetic segmentation and recoding in the beginning reader. In A.S. Reber & D. Scarborough (Eds.), *Toward a psychology of reading.* Hillsdale, NJ: Erlbaum.
O'Dougherty, M., Nuechterlein, K.H., & Drew, B. (1984). Hyperactive and hypoxic children. Signal detection, sustained attention, and behavior. *Journal of Abnormal Psychology, 73*, 178-191.
Richardson, E. (1984). The impact of phonemic processing instruction on the reading achievement of reading-disabled children. *Annals of the New York Academy of Sciences, 433*, 97-118.

Richardson, E., & DiBenedetto, B. (1977). Transfer effects of a phonic decoding model: A review. *Reading Improvement*, *14*, 239-247.

Richardson, E., Foss, D., & DiBenedetto, B. (in press). Specific decoding skill deficits in diagnosing reading problems. *Annals of Dyslexia.*

Satin, M.S., Winsberg, B.G., Monetti, B.A., Sverd, J., & Foss, D.A. (1985). A general population screen for Attention Deficit Disorder with Hyperactivity. *Journal of the American Academy of Child Psychiatry*, *24*, 756-764.

Swanson, H.L. (1980). Auditory and visual vigilance in normal and learning-disabled readers. *Learning Disabilities Quarterly*, *3*, 71-78.

Swanson, H.L. (1981). Vigilance deficit in learning-disabled children: A signal detection analysis. *Journal of Child Psychology and Psychiatry*, *22*, 393-399.

Swanson, H.L. (1983). A developmental study of vigilance in learning-disabled and nondisabled children. *Journal of Abnormal Child Psychology*, *11*, 415-428.

Symmes, J.S., & Rapoport, J.L. (1972). Unexpected reading failure. *American Journal of Orthopsychiatry*, *42*, 82-91.

Vellutino, F.R. (1979). *Dyslexia: Theory and research*. Cambridge, MA: The MIT Press.

SELECTED READINGS

Achenbach, T. M. (1980). The DSM-III in light of empirical research on the classification of child psychopathology. *Journal of the American Academy of Child Psychiatry*, *19*, 395-412.

Achenbach, T. M. (1982). Assessment and taxonomy of children's behavior disorders. In B. Lahey & A. Kazdin (Eds.), *Advances in Clinical Child Psychology* (Vol. 5). New York: Plenum.

Barkley, R. (1976). Predicting the response of hyperactive children to stimulant drugs: A review. *Journal of Abnormal Child Psychology*, *4*, 327-348. Reprinted in B. Lahey (Ed.), *Behavior therapy with hyperactive and learning disabled children*. New York: Oxford University Press, 1979.

Barkley, R. A. (1982). Guidelines for defining hyperactivity in children (Attention Deficit Disorder with Hyperactivity). In B. Lahey & A. Kazdin (Eds.), *Advances in clinical child psychology* (Vol. 5, pp. 137-180). New York: Plenum.

Campbell, S. B. (1976). Hyperactivity: Course and treatment. In A. Davids (Ed.), *Child personality and psychopathology* (Vol. 3). New York: John Wiley and Sons.

Campbell, S. B. (1985). Early identification and follow-up of parent-referred "hyperactive" toddlers. In L. M. Bloomingdale (Ed.), *Attention Deficit Disorder: Emerging trends* (Vol. II). New York: Spectrum.

Cantwell, D. P. (1980). Clinician's guide to the use of stimulant medication for the psychiatric disorders of childhood. *Developmental and Behavioral Pediatrics*, *1*(3), 133-140.

Cantwell, D. P. (1980). Drugs and medical intervention. In E. D. Rie & H. E. Rie (Eds.), *Handbook of minimal brain dysfunctions: A critical view* (pp. 590-617). New York: John Wiley and Sons.

Conners, C. K., Denhoff, E., Millichap, J. G., & O'Leary, S. C. (1978). Hyperkinesis therapy: Posing therapeutic options to parents. *Patient Care*, *7*, 94-154.

© 1987 by The Haworth Press, Inc. All rights reserved.

Conners, C. K. (1985). The computerized continuous performance test. Psychopharmacology Bulletin, 21(4), 891-892.

Goyette, C. H., Conners, C. K., Ulrich, R. F. (1978). Normative data on revised Conners' parent and teacher rating scales. Journal of Abnormal Child Psychology, 6(2), 221-236.

McClure, F. D., & Gordon, M. (1985). Performance of disturbed hyperactive and nonhyperactive children on an objective measure of hyperactivity. Journal of Abnormal Child Psychology, 12, 561-572.

Pelham, W. E., & Murphy, H. A. (1986). Behavioral and pharmacological treatment of attention deficit and conduct disorders. In M. Hersen (Ed.), *Pharmacological and behavioral treatment: An integrative approach.* New York: John Wiley and Sons.

Richardson, E., Winsberg, B. G., & Bialer, I. (1973). Assessment of two methods of teaching phonic skills to neuropsychiatrically impaired children. Journal of Learning Disabilities, 6, 628-635.

Satterfield, J. H., Cantwell, D. P., Satterfield, B. T. (1979). Multimodality treatment: A two-year study of 61 hyperactive boys. Archives of General Psychiatry, 36, 965-974.

Wells, K. C., Conners, C. K., Imber, L., & Delameter, A. (1981). Use of single-subject methodology in clinical decision-making with a hyperactive child on the psychiatric inpatient unit. Behavioral Assessment, 3, 359-369.

Documents and Journal Articles from the ERIC Database

ERIC DOCUMENTS

Brown, E. D. (1983). Hyperactivity: Associated behaviors in the social-ecological context of preschools. (ERIC Document Reproduction Service No. ED 226 559)

Brown, E. D. (1984). Social interactions of preschool hyperactive boys. (ERIC Document Reproduction Service No. 235 924)

Milich, R. (1984). Two year stability and validity of playroom observations of hyperactivity. (ERIC Document Reproduction Service No. ED 239 442)

Zentall, S. (1982). Language and activity of hyperactive and comparison children during listening tasks. (ERIC Document Reproduction Service No. ED 216 483)

JOURNAL ARTICLES

Cohen, N. J., & Minde, K. (1983). The "hyperactive syndrome" in kindergarten children: Comparison of children with pervasive and situational symptoms. *Child Psychology and Psychiatry and Applied Disciplines*, *24*(3), 443-455. (ERIC Document Reproduction Service No. EJ 284 389)

Mash, E. J., & Johnston, C. (1983). The prediction of mothers' behavior with their hyperactive children during play and task situations. *Child & Family Behavior Therapy*, *5*(2), 1-14. (ERIC Document Reproduction Service No. EJ 292 552)

Pollard, S. (1983, Winter). The effects of parent training and ritalin on the parent-child interactions of hyperactive boys. *Child & Family Behavior Therapy*, *5*(4), 51-69. (ERIC Document Reproduction Service No. EJ 300 840)

JOURNAL ARTICLES are cited and annotated in the monthly publication *Current Index to Journals in Education* (CIJE). These articles may be read in periodicals obtained in libraries or through subscription. Selected article reprints are available from University Microfilms International, Article Reprint Department, 300 N. Zeeb Road, Ann Arbor, MI 48106. Please see the most recent issue of *CIJE* for UMI ordering details.

Further information about the ERIC network and services of the ERIC Clearinghouse on Handicapped and Gifted Children are available from ERIC/EC Information Services, 1920 Association Drive, Reston, VA 22091 (703/620-3660).

Subject Index

Abbreviated Teacher Questionnaire, value of, 49-50
Academic Intervention
 guidelines, 165
 Integrated Skills Method, 155-158
 medication, value of, 159
 studies on, 159-165
 therapeutic milieu, 158-159
 See also Developmental Reading Disorder, Psychological tests, Treatment procedures
Academic skills
 case examples, 61-63,92-93
 prognosis, 90
 young adults, 90
Actometer. *See also* Structured Observation of Academic and Play Settings
Achievements/underachievements
 personality traits, 108-109
 sex differences, 108-109
Affective disorder, 8-9
Aggressive behavior, 24
 boys, 69
 case example, 91-93
 diagnosis of, 24
 prognosis, 89,91,97
 See also Child Behavior Checklist, Structured Observation of Academic and Play Settings
Anger, 60
Antisocial behavior
 case example, 96-98
 See also Social Development
Anxiety
 diagnosis, 22-23,36-37,49,60
 See also Teacher Rating Scales
Assessments
 criteria, 78
 interviews, 79
 See also Diagnosis, DSM-111/DSM-111-R

Attention Deficit Disorder, *xi*
 boys, 69
 case examples, 16,89-98,91-93
 classification, 5-6
 computerized assessment, 53-55
 diagnosis of, 5,22-23,33,41,75
 follow-up studies, 90
 issues, developmental, 10
 issues, gender, 10-11
 pharmacological intervention, 101
 prognosis, 89
 relationships, peer, 10
 See also Attention Deficit Disorder with Hyperactivity, Diagnosis, DSM-111/DSM-111-R, Hyperactive Syndrome, Teacher Rating Scale
Attention Deficit Disorder, Residual Type
 diagnosis, 6-11
Attention Deficit Disorder with Hyperactivity
 boys, 69
 case examples, 62,78-83,85,91-92, 94-96
 computerized assessment, 63
 Developmental Reading Disorder relation to, 153
 diagnosis of, 6-11,36,53,59,63, 65-67,71-72,74
 diagnostic issues, 65
 measure of, 35-36
 See also Developmental Reading Disorder, Preschoolers, Structured Observation of Academic and Play Settings, Treatment Procedures
Attention Deficit Disorder without Hyperactivity
 diagnosis of, 6-8
 case example, 61
Attention span

© 1987 by The Haworth Press, Inc. All rights reserved.

age considerations, 66
case example, 89-98
parent observations, 66
pharmacological intervention, 66
teacher observations, 66
See also Pharmacological Interventions

Behavior Therapy
behavior contracts, 113
case example, 119-120
home intervention, 113
operant reinforcement, 113
parent training, 113-115
peer intervention, 113,117
relaxation training, 113
token economics, 113
school intervention, 113,115-116
social skills training, 116-117
techniques, 111
value of, 111-112,117-119
See also Prognosis, Treatment procedures
Behavior rating scales, 14-15
See also Diagnosis
Bender-Gestalt, 94
See also Psychological tests
Boys
aggression, 65,69,71-72
hyperactivity, 72-73
See also Attention Deficit Disorder with Hyperactivity
Brain dysfunction, minimal, 9

Chemical abuse
young adults, 91
Child Behavior checklist, 19,20-30
child behavior profile, 21-23
components, 24-29
DSM-111/DSM-111-R, relation to, 22-23
evaluation of, 29-30
intake procedures, 23
interviews, 22-24
parent participation, 20-21,23-24
reassessments, 29
school problems, 27
social competence, 24
syndromes, 22-24

Teacher's Report Form, 27,29
See also Parent Rating Scale
Child Behavior Profile. *See also* Child Behavior Checklist
Classroom activity
evaluation of, 35-36
See also Teacher Rating Scale
Clinicians
teacher relationships, 33-34
Cognitive Behavior Therapy
attribution intervention, 126-129
case example, 128
clinical issues, 132-137
guidelines for application, 137-140
multipurpose approach, 130-132
origins, 123-125
self-direction, 129
self-instruction, 125,129
self-monitoring, 124-125
See also Treatment procedures
Competence, 21-22
Computerized assessment, 53-55,58
See also Gordon Diagnostic System
Conduct disorder, 4,22-23,36,49
diagnosis, 36-37,41,49
See also Teacher Rating Scales
Continuous Performance Test, 54
See also Gordon Diagnostic System
Cocaine, 95

Decoding Skills Test, 159
See also Academic Intervention, Developmental Reading Disorder
Delinquent behavior, 24,27,29
diagnosis of, 24
young adults, 91
See also Child Behavior Checklist, Social development
Depression, 20-24,27-29,58
diagnosis of, 24,49
prognosis, 90
See also Child Behavior Checklist
Developmental Reading Disorder
Attention Deficit Disorder, relation to, 153-154
diagnosis of, 159
Integrated Skills Method, 155-158
methylphenidate, value of, 159

Subject Index

therapeutic milieu, 158-159
treatment results, 160-162
See also Academic Intervention
Diagnosis
assessment procedures, 75,86
Behavior Rating Scale, 14-15
Child Behavior Checklist, 20-30
clinical observations, 71
criteria, 53
criteria, age, 9
criteria, duration, 9
criteria, exclusion, 9
criteria, inclusion, 9
developmental issues, 10,75-77
evaluation process, 11-16
interview process, 11-14
issues, 33
multiaxial approach, 6,19,29-30,49
multi-informant rating scales, 14-15
Parent Rating Scale, 14-15
See also Child Behavior Checklist
parental role, 76
Teacher Rating Scale, 14-15,33-50
Teacher Report Form. *See also* Child Behavior Checklist
young adults, 90
See also Attention Deficit Disorder, Behavior Rating Scales, DSM-111/DSM-111-R, Gordon Diagnostic System, Parent Rating Scale, Teacher Rating Scale, Treatment procedures
DSM-111/DSM-111-R, 33-34
case examples, 16,59
classification, 6
diagnostic criteria, 5-11,19,61-62, 65,75
origin, 6
peer relationships, 10
teacher evaluations, 33
Teacher Rating Scale, relation to, 40
syndromes, 22-23
See also Attention Deficit Disorder, Attention Deficit Disorder with Hyperactivity, Attention Deficit Disorder without Hyperactivity, Child Behavior Checklist, Hyperactive syndrome

Figure drawings. *See also* Psychological tests
Follow-up profiles, 29
See also Child Behavior Checklist, Prognosis

Gates-MacGinitie, 159
See also Psychological tests
Gordon Diagnostic System
brain damage, diagnosis of, 57
case example, 58-63
classroom behavior, diagnosis of, 57
Delay Task, 56,58-59,61
Distractability Task, 56
preschoolers, diagnosis of, 57
value of, 56-58
Vigilance Task, 56,58-59,61-62

Hallucinogens, 95
Hashish, 95
Heroin, 95
Hollingshead and Redlich Two-Factor Index, 92
Home visits, 79
See also Interviews
Hyperactive syndrome, 6-9,13-14,19,24, 27,58,65
behavior, pervasive, 81
behavior, situation specific, 81
boys, 69
case examples, 24-29,58-60,78-86
characteristics, 65
definition, *xi*
diagnosis of, *ix*,24,27,29,36-37, 40-41,49
diagnostic indicators, 79
diagnostic issues, 1-2
DSM-111/DSM-111-R, *ix*,1
prognosis, *x*,2,91
prognosis for preschoolers, 75-77
research, *ix*
studies, 77-78
treatments, *x*,1-2
See also Attention Deficit Disorder with Hyperactivity, Child Behavior Checklist, DSM-111/DSM-111-R, Structured Observation of Academic and

Play Settings, Teacher Rating Scales, Treatment procedures
Hyperkinesia, 9,90,94

Impulsivity, 6-9,13-14,40,58,63,65,143
  absence of, 61
  case examples, 58-60,61-63,71, 89-98,92-96
  definition, 8
  diagnosis of, 35-36,40
  medications, 101
  prognosis, 90
  See also Attention Deficit Disorder with Hyperactivity, Pharmacological Intervention, Teacher Rating Scales, Treatment procedures
Inattention, 4,6-9,13-14,24,27,29,65,143
  case examples, 59-60,61-63
  diagnosis of, 40
  See also Attention Deficit Disorder without Hyperactivity, DSM-111/DSM-111-R, Hyperactive Syndrome, Teacher Rating Scale
Intake procedures, 23-24
  See also Child Behavior Checklist
Intervention, 12
  See also Treatment procedures
Interviews, 11-14,22-23,29-30
  children, 60
  clinical, 61
  considerations, 11-14
  home, 80,83-84
  parents, 60,70,72,78-86,90
  SADS-L, 92-93,95
  teachers, 60
  young adults, 90,92
  See also Child Behavior Checklist, Diagnosis
Iowa Project, 89-90
  See also Attention Deficit Disorder

Kagan Matching Familiar Figures Test, 92-93,95
  See also Psychological tests

Marijuana, 93,95
Medical exam, 29-30

Medications. See also Pharmacological Intervention
Mental retardation, 8-9
Methadone, 95
  See also Pharmacological Intervention
Methylphenidate. See also Pharmacological Intervention
Multi-Informant Rating Scale. See also Diagnosis
Motor behaviors
  diagnosis of, 35-36
  See also Teacher Rating Scales

Observations, 35-36
  direct observations, 29-30
  home, 81
  observers, 30
  teachers, 81
  See also Diagnosis, Interviews, Teacher Rating Scales
Occupational development
  employment, 94
  impairment of, 8
Overactivity, 27,29,66
  diagnosis of, 27
  prognosis, 90
  See also DSM-111/DSM-111-R, Hyperactive syndrome

Parents, 143
  assessment, role in, 23-24
  behavior therapy training, 113
  diagnosis, participation in, 19,27,29,33
  influence on behavior, 85,87
  influence on ADD behavior, 145
  See also Behavior Therapy, Child Behavior Checklist, Diagnosis, Interviews, Parent Training, Pharmacological Intervention, Treatment procedures
Parent Rating Scales, 14-15,20-21
  diagnostic process, 19
  scores on, 69
  See also Child Behavior Checklist, Diagnosis
Parent Training
  behavior management skills, 143
  case example, 149-150

Subject Index

group training, 143
guidelines, 147-149
value of, 145-147
See also Treatment procedures
Peabody Individual Achievement Test, 159
See also Psychological tests
Peer relationships, 10
case examples, 61-63,72,94-96
prognosis, 90
Personality disorders, 8
Pharmacological Intervention
assessments, clinical, 104-110
case examples, 61-63,72,90,92-93, 105-108,119-120
clinical issues, 99,100-104
Developmental Reading Disorders, relation to, 159,162-165
dextroamphetamine, 93-94,100
home behavior, relation to, 112
imipramine, 93
methylphenidate, 62,82,90,93,100, 153,159,162
notriptyline, 93
pemoline, 100
response to Structured Observation of Academic and Play Settings, 73
stimulants, 59,61,99,112
See also Prognosis, Treatment procedure
Peabody Individual Achievement Test. *See also* Psychological tests
Play, free
observations during, 78
See also Structured Observation of Academic and Play Settings
Playroom observation procedure. *See also* Structured Observation of Academic and Play Settings
Preschoolers, hyperactive
case examples, 78-86
diagnostic indicators, 79-80,84
parental role. *See also* Treatment procedures
prognosis, 75-87
See also Hyperactive syndrome, Parents
Prognosis, 87
Attention Deficit Disorder, 90

behavior therapy, use of, 124
case examples, 89-98
developmental issues, 87
hyperactivity, 124
individual differences, based on, 130-131
medications, use of, 124
preschoolers, 75-87
Psychological tests
Bender-Gestalt, 60,94
Figure drawings, 60
Gates-MacGinitie, 159
Kagan Matching Familiar Figures Test, 92
Peabody Individual Achievement Test, 159
Rorschach, 60
scores on, 69
Stanford-Binet Scale of Intelligence, 78
Thematic Apperception Test, 60
Wechsler Adult Intelligence Scale, 92-93
Wechsler Intelligence Scale for Children, 60,94
Wide Range Achievement Test, 60,92-93
Psychometrics, 29-30
Psychopathology
case example, 62
See also Abbreviated Teacher Questionnaire
Psychostimulants. *See also* Pharmacological intervention

Questionnaires, 66
patient, 90
teacher questionnaires, 51
See also Abbreviated Teacher Questionnaire

Reading Grade Equivalent, 160
Reassessments, 29
See also Child Behavior Checklist
Restricted Academic Period. *See also* Structured Observation of Academic and Play Settings
Rorschach, 60
See also Psychological tests

SADS-L. *See also* Interviews
School problems, 27
    See also Child Behavior Checklist
Schizophrenia, 8-9
Self-esteem, 90
Social competence, 24
    See also Child Behavior Checklist
Social development, 10,143
    antisocial behavior, 89,91,93,96
    diagnosis of, 27
    impairment of, 8
    problems with, 144-145
    See also Attention Deficit Disorder
Stanford-Binet Scale of Intelligence, 78,81,84
    See also Psychological tests
Structured Observation of Academic and Play Settings (SOAPS)
    actometer, 67
    aggressive boys, 69,71-72
    behaviors coded, 68
    case examples, 69-73
    free play periods, 67-68
    medication, relation to, 72-73
    motor activity, measure of, 67
    origin, 66
    playroom equipment, 67
    population statistics, 69-73
    Restricted Academic Period, 67-69
    studies, conducted on, 68
    toys, 67
    value of, 66,74
Suicide
    case example, 61-63

Teachers, 19
    diagnosis, participation in, 24,27, 29-30,33-34
Teacher Rating Scales, 14-15,33-51,159
    anxiety, diagnosis of, 36-39,49
    conduct disorders, diagnosis of, 36-37
    cost-effectiveness, 35-36
    depression, diagnosis of, 49
    direct observation, 35-36
    halo effect, 49
    hyperactivity, diagnosis of, 36,40
    inattentiveness, diagnosis of, 36

impulsivity, diagnosis of, 40
limitations of, 49
reliability/validity, 35,41
response, evaluation of, 40-49
scores on, 69
scoring of, 41-49
value of, 34
See also Abbreviated Teacher Questionnaire, Diagnosis, DSM-111/DSM-111-R
Teacher's Report Form, 27
    See also Child Behavior Checklist
Thematic Apperception Test. *See also* Psychological tests
Therapeutic goals, 29
    See also Child Behavior Checklist, Treatment procedures
Treatment
    Academic Therapy. *See also* Academic Intervention
    behavior management, 78,82
    Behavior Therapy. *See also* Behavior Therapy
    Cognitive Behavior Therapy. *See also* Cognitive Behavior Therapy
    intervention, early, 87
    medications. *See also* Pharmacological Intervention
    parent training, 78,81-82,86
        See also Parent Training
    strategies, 61-62
    See also Developmental Reading Disorder

Unpopular scale, 27
    See also Teacher's Report Form

Wechsler Adult Intelligence Scale, 92-94
    See also Psychological tests
Wechsler Intelligence Scale for Children, 94
    See also Psychological tests
Wechsler Intelligence Scale for Children—Revised, 60
    See also Psychological tests
Wide Range Achievement Test, 92-94
    See also Psychological tests